Perfect Symmetry

By Evan Olsen

RoseDog ❧ Books

PITTSBURGH, PENNSYLVANIA 15238

RoseDog Books
585 Alpha Drive
Suite 103
Pittsburgh, PA 15238
Visit our website at www.rosedogbookstore.com

ISBN: 978-1-6442-6706-6
eISBN: 978-1-6442-6728-8

$$\alpha=\Omega=\Delta=\Psi$$

Perfect Symmetry

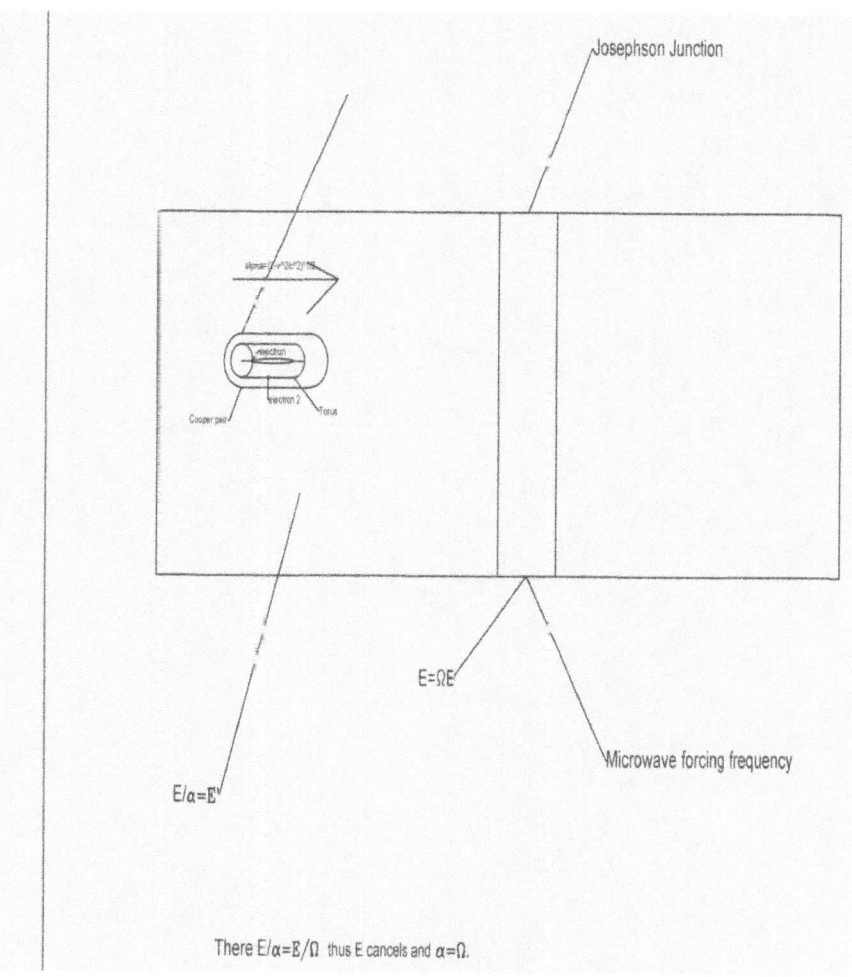

Josephson Junction

$$E=\Omega E'$$

Microwave forcing frequency

$$E/\alpha=E'$$

There $E/\alpha=E/\Omega$ thus E cancels and $\alpha=\Omega$.

Two quantum entangled electrons form a torus which forms the basis for space-time that inelastic collisions can propagate faster then light.

Table of Contents

Introduction.

If in the fifth book the quantum entangled cooper pairs are indeed colliding in elastically faster than the speed of light and thus the constant is replaced with the real equation for dependence of mass as $E=m((u-u')/omega)^2$ thus proving that there is no dependence on the speed of light for the real configuration of space time and thus Einstein's equation for the dependence of mass and the constant of light is indeed is not the total picture and makes one wonder as to the importance of superconductivity and its relation to the struggle in world war 2 when the Germans had the Meissner effect and Hitler's coming to power in 1933 and the discovery of the Josephson effect in 1962 culminating with the Cubin missile crises. Father more if neural networks can be quantum entangled the question arises as to the importance of this struggle and what it really means and the deception that was the twentieth century. The deception that aliens have no influence on our culture and society. Rather it's not that Einstein's equation is wrong since it has been experimentally verified it's just it's a subset in a much larger universe just as classical mechanics was in Einstein's universe.

But first it is ultimately important to understand and go through what the first step of the equation means the first proof is proving that space time was torus and it can be shown with the transformation of the square to cylinder to torus transform that the inertial system of reference can indeed be equated to omega to the circle map equation and furthermore that this can be equated right at the junction point of the Josephson effect and that this is equivalent to the third equation delta that things are colliding faster than the speed of light through quantum tunneling. The purpose of this book is to give the description towards the development that all viruses are transforming through this Mobius transform and that I fear one of the greatest viruses that plagues mankind is transforming through this transformation. Further the Mobius transform connects two systems of reference or two space times and thus is teleportation. Knowledge of the true structure of DNA of which we are deciphering as being bound torus's above and below the plane of the aromatic pi bonds that are colliding in elastically that can be faster than the speed of light to be able to give DNA its energy to persist in the universe as a mass energy conversion device. The question now relates to the linking of these pi bonds between adjacent strands and hence the purpose of these books the Mobius transform and the fourth equation. DNA is a factor of proportionality between order and disorder and this comes from the quantum entangled pi bonds that form a relativistic curve right at the junction point proving that the first equation the Lorentz transformation is mathematically equivalent to quantum mechanics at the omega equation and this equation is for quantum tunnelling and combines chaos theory, relativity and quantum mechanics. Current colliders are completely out of date and based on the equation of $E=Mc^2$. To just let people know I am not trying to rewrite history I am just stating scientific facts and there mathematical certainties that will liberate the current population of which there is one race the human race. DNA is the junction point and things collide faster than the speed of light and Einstein is wrong about the absolute fastest speed because space-time is a torus but can be factored into the Einstein's field equation because the torus's equate to an equilateral which according to Tensors Differential Forms and variational Principles By David Lovelock and Hanno Rund forms the curve in Euclidean space that forms the equilateral triangle and can be defined by the Euclidean metric of the tangent space and thus $1/p=d\phi/db$.

Furthermore it has been discovered that in Josephson junctions it's possible to make something called thermal rectifiers that reverse the flow of entropy across the junction when changing the direction of one the parameters of the winding equation as my theory predicts this is time travel and this means the micro world of quantum tunneling can be brought to the world of relativistic physics and the very large like spaceships or people. My father always said you can't prove things with philosophy because it is not valid without an equation and the merits of this equation can be proven by itself. This is a valid equation and I have human rights and I believe in humanity and I believe in technology that will liberate us. Remember the real reason one would discount the alpha equal omega equation is there fear of alien intelligences traveling here for if one could posit that astronomical objects could quantum tunnel faster then light it would mean it would not take aliens thousands of years to get here hindered by the speed of light. This would have religious consequences; that DNA is centered around this circle map or the space time continuum life must be abundant everywhere. This is both good and bad as one can posit that if there are good aliens there must be bad aliens. However we must approach all aliens with the message of peace but be ready with the equations that I have deciphered here to be able to adequetely defend ourselves in the event that they are hostile. Research on this technology MUST happen immediately and human beings need to sober up and accept the reality. There is an excellent book written called Aztec; in it the main character has adventures all throughout the book but he is the first to see the European ships in the sea. He recognizes they are more advanced and tells his leaders of the threat they may pose and the incoming doom that may happen, they don't listen!!! There culture is wiped out and destroyed. Well if all the video proof and audio proof of alien spacecraft that weve been seeing if all the crop circles and all the stuff that proves beings are more intelligent then us or more advanced is true then we must prepare humanity. Further theories like string theory that have no observable experiments and is pure speculation and does not have repeatable experiments can not be taken as essential scientific truths. It not even a theory because it doesn't explain anything in reality it's a hypothesis that's never been proven. I have tons of experiments that proves my theory real and makes predictions that come true and is repeatable.

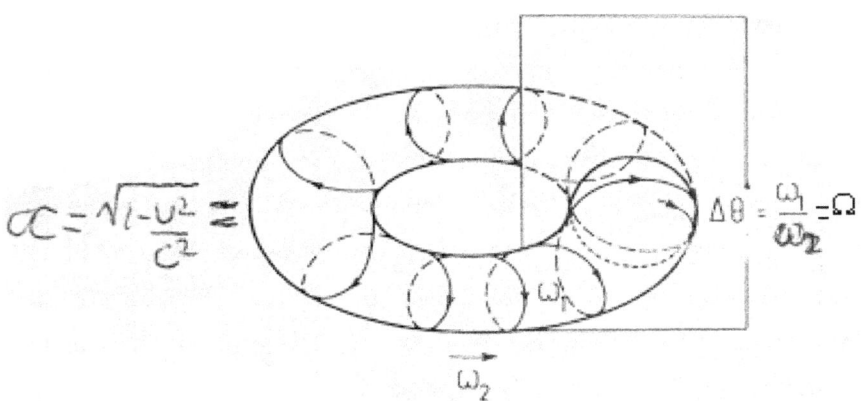

Evan Olsen

The Harmonist

Chapter 1

Harmony Theory

Further on Space Time and DNA

(1-v^2/c^2)^1/2=hw1/hw2=(1-T1/T2)^1/2

Evan Olsen B.S.c Physics and B.S.c Psychology

University of Alberta

Abstract.

The sine circle map $\theta n+1=\theta n+\Omega-(k/2\pi)\sin(2\pi\theta n)$ is investigated from a relativistic perspective. The Ω **(frequency ratio)** or average shift of the angle theta or winding number is conjectured to be a **lorentz transformation (alpha) or** $\alpha=\Omega$. This has ramifications for space-time theory and because of its relation to chaos theory also has ramifications to DNA. The winding number in the sine circle map is investigated experimentally by the Josephson junction in a microwave field. DNA is considered to be a strange attractor and its structure is related to this new theory of space-time. Space-time is an oscillator and so is matter, and thus space-time is the causation of life through chaos.

The motivation behind this paper is to see if space-time can be described in a new way as to account for the possibility of life. The concept of Alpha equal Omega equal delta brings Harmony to three major physical theories, relativity, quantum theory and chaos theory. The scientific belief that harmony can describe the universe and life in it! The belief that DNA mirrors something fundamental and basic about our universe and that fundamental thing is space-time! The new model of space-time allows for the possibility of time travel and mass-energy conversion. As well the idea that DNA is a strange attractor could lead to cures for cancer.

The sine circle map is considered to be one of the most important equations governing the dynamics of chaos. Its equation **θn+1=θn+Ω-(k/2π)sin(2πθn) Mod 1,** represent motion on a unit torus (fig1) where the frequency ratio is the winding number or Omega value.

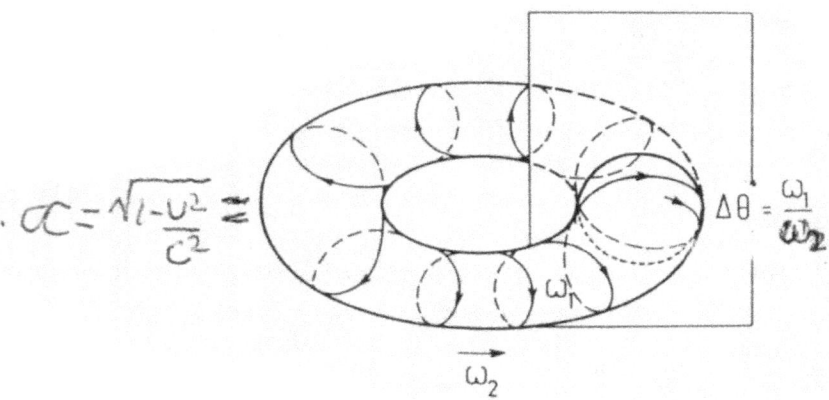

Fig. 1 The winding number on the torus equals a lorentz contraction

The unperturbed circle map is just **θn+1=θn+ Ω Mod 1** and its motion is periodic if the frequency ratio p/q is rational. The motion on the unit torus is quasiperiodic if p/q is irrational and the motion eventually covers the whole torus[3]. This sine circle map is studied comparing values of omega to the coupling parameter K, where the break up of periodic motion happens to chaotic motion where K is increasing to greater than 1 fig 2.

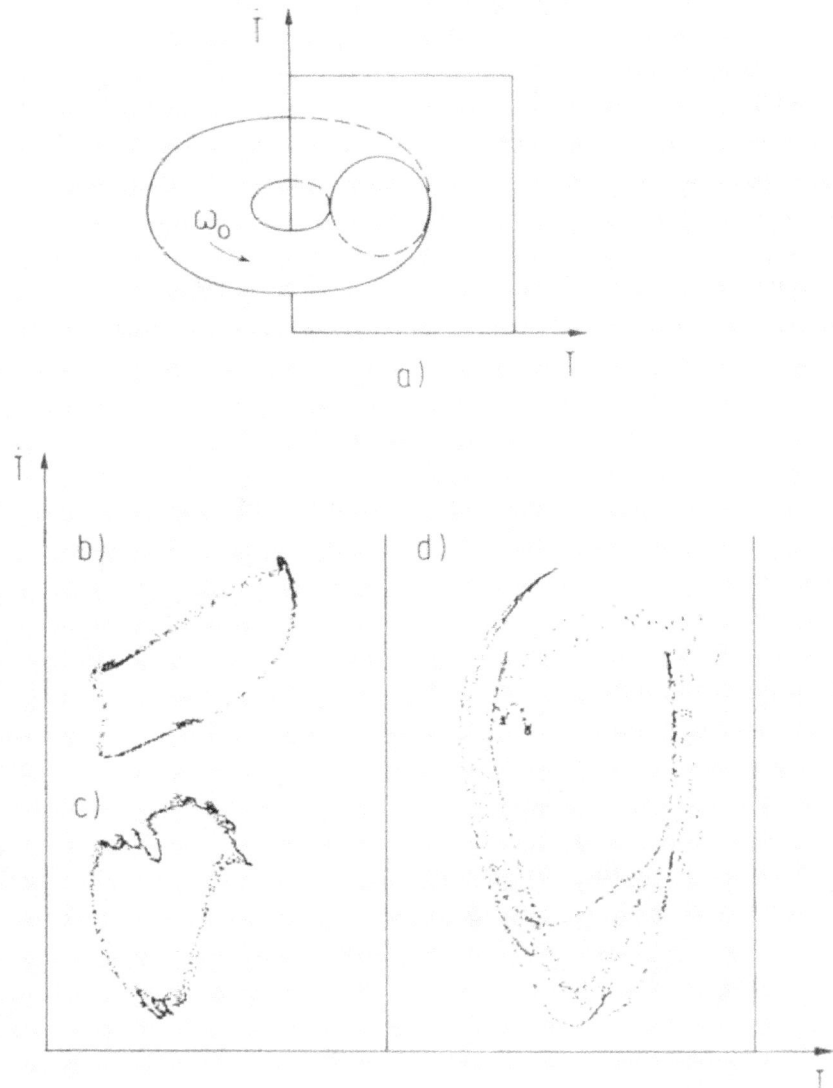

Fig2. Due to the parameter K causes the formation of a strange attractor from the torus

The Lorentz contraction (v=velocity of object c=velocity of light) was first purposed by Lorentz through the ether as the result of a supposed effect of motion on the electromagnetic forces between the particles making up the body. Every body which has the velocity v with respect to the ether contracts in the direction of motion by the factor alpha = $\sqrt{1-v^2/c^2}$. Soon after Einstein came up the idea of relative simultaneity and ruled out the possibility of absolute simultaneity. This is where the idea of a Lorentz contraction was created to connect two systems of reference in space-time that are relatively simultaneous. The Lorentz contraction or transformation is called the factor of proportionality and connects two systems of reference moving with respect to each other. The concept of space-time was invented to explain this Lorentz transformation phenomenon and from the conception of a space-ct axis the transformation from one system of reference to another was born in space-time[1].

However there is another interpretation that deals with the idea of rotation. Since time is measured by rotation then all things in the universe depend on rotation. Time is measured by the rotation of the earth (day), and the linear interpretation of time is in intervals where by the completion of a rotation is of sometime understood. If space-time is connected then space should also be governed by rotation. If two events in space-time are governed by the connection between two systems of reference having different space-ct axis then it is possible to describe two rotating systems of reference with respect to each other. The two rotating systems represent the two systems with space-ct axis. This is only true if the relationship between the systems of reference is relatively simultaneous. This is where the sine circle map comes into play. The omega value is a factor of proportionality between two simultaneously moving systems of reference. The two systems W1 and W2 represent the motion of the minor rotation over the major rotation w1/w2. If the two systems are moving with respect to each other simultaneously then the frequency ratio must be a Lorentz contraction. The factor alpha which is the factor of proportionality between two systems of reference is the same as omega, which connects two rotating systems of reference on the torus or sine circle map.

Furthermore the Lorentz contraction is for straight line motion and if the winding number on a torus is a straight line on curved space, then alpha or a Lorentz contraction is equal to omega or

1) $\alpha = \sqrt{1-v^2/c^2)}=\Omega=w1/w2$

Thus $\alpha=\Omega$ because the two motions are both straight lines. Omega is a straight line governed by the two rotating systems on the torus and the winding number is a geodesic that determines the periodicity and is the Lorentz contraction straight line motion in regular Euclidean space. According to Max Born's book on Einstein's Theory of Relativity "...it follows that the geodetic (geodesic) lines must correspond exactly to those physical phenomena which are represented by straight lines- namely, rays of light and motions of inertia"[1]. In fact if alpha were not equal to omega this would contradict Einstein's General Theory of Relativity. Plus the torus can be generated from a square or flat Euclidean space by first folding the space into a cylinder and then gluing the two ends of the cylinder together. Thus the straight lines on the square Euclidean space form the winding number of the map. Hence the motions are the same.

Proof of this comes from the next derivation:

Algebraic Proof for Alpha = Omega

$\alpha = \Omega$

$\theta > 45°$ so $\frac{y}{x} > 1$ — only true for these values

lets use

$\tan 50° = 1.192$

$\gamma = \frac{1}{\left(1 - \frac{v^2}{c^2}\right)^{1/2}}$

$1.192 = \frac{y}{x} = \frac{1}{\gamma} \frac{\Delta y'}{\Delta x'} = \frac{1}{\gamma} \tan \alpha'$

$1.192\, \gamma = \frac{\Delta y'}{\Delta x'}$

Lets move to Translation   moving frame to torus

radius x'

$\frac{1}{\gamma} = \Omega \gamma$

$r_1 < r_2$

radius of $\Delta y'$

$0 < \Omega < 1$

$0 < \frac{\Delta x'}{\Delta y'} < 1$

$1.192\, \gamma = \frac{\Delta x'}{\Delta y'} = \frac{r_1\, \theta_1}{r_2\, \theta_2} = ?$

$.84\, \Omega = \frac{\Delta x'}{\Delta y'} = \frac{r_1\, \theta_1}{r_2\, \theta_2} = \frac{r_1}{r_2} \Omega$

$1 > .84 = \frac{r_1}{r_2}$ aspect ratio is correct

This is only true if $\Omega = \frac{\theta_1}{\theta_2}$ for ring torus radius

$\frac{1}{\gamma} = \alpha = \Omega = \frac{\theta_1}{\theta_2} = \sqrt{1 - \left(\frac{v^2}{c^2}\right)}$

Fig 3 Alpha=Omega algebraic proof

If this is true one should expect the value of omega for the sine circle map to be analogous to alpha. Alpha can be expressed as a frequency

6

ratio and it can be expressed as a change in angle between two systems just like omega which is the change in the angle theta per iteration of the circle map. In the previous diagram we see that the Lorentz transform for the moving frame becomes the torus with the angle ratio which is the frequency ratio. As well omega is bounded from zero to one just like the Lorentz contraction. Yet this winding number can also have negative frequency allowing for time dilations to the past. Alpha can be expressed in action angle variables as $\omega n = \partial H/\partial J_n$ the frequencies the partial derivative of the Hamiltonian with respect to the action J. So

2) $(\partial H/\partial J1)/(\partial H/\partial J2) = w1/w2 = \alpha = \Omega$

Is the Harmonic Oscillator on the torus!

What $\alpha = \Omega$ means for space-time is that space-time is an oscillator or torus!

This means the contraction itself is an oscillation and one may produce a Lorentz contraction without incredible speeds of v just by the frequency ratio omega. If the object takes on the winding number and winds on the torus it should by this theory produce a Lorentz contraction. If this is true then mass objects are space-time configurations and these configurations should be oscillators themselves. As shown

3) $E'/(\sqrt{1-v^2/c^2}) = E'/\alpha = E'/\Omega = E$

This equation states that the energy at rest is a fraction of the energy of motion by a factor of proportionality. As velocity increases the energy increases. If $\alpha = \Omega$ then the frequency ratio of the torus becomes smaller (w1/w2) then the energy should increase the same as for a Lorentz contraction. Time dilation is another relationship

4) $\quad T\sqrt{(1-v^2/c^2)}=T(\alpha)=T(\Omega)=T'$

Or that moving clocks run slower than stationary clocks. It is also the case that if omega is not dependent on velocity but is equal to a Lorenz contraction then a negative frequency ratio is possible and time travel is possible.

So if alpha is equal to omega then these relationships should be proven true by an experiment. This experiment will be talked about later in results and discussion. For now let us consider the space-time double helical structure of DNA. The realm of chaos governs this new theory of DNA. It is through this chaos that the harmony can be seen. This theory of DNA states that DNA is the main oscillator for the cell. By this definition then, living structures can be regarded mathematically as disequilibrium/ equilibrium structures interacting with a chaotic universe to produce harmony as an oscillator. The question is what type of oscillation in physics are we talking about? If DNA is considered to be dependent on initial conditions when in motion for super coiling; then its motion can be considered to be stochastic and impossible to predict mathematically. This type of oscillator is known to be a strange attractor! This is an oscillator that is periodic (recurrent) but never repeats the same trajectory twice as shown in the figure of a Lorenz attractor fig (3).

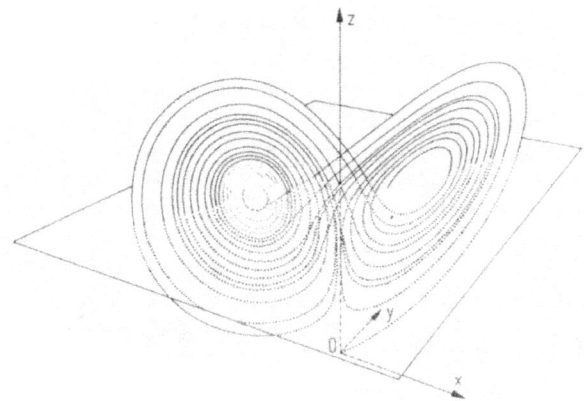

Fig4. The Harmony Structure in chaos, the strange attractor of Lorenz.

This is where chaos becomes harmonious; the fact that DNA is an oscillator that programs life then life is an expression of the non-linearity inherent in space-time governing the evolution of the organism. Thus life is an expression of the repeating modes of chaos to find equilibrium in chaotic flow. How does DNA relate to space-time?

Since DNA is a structure that is winding like the Lorentz contraction of the torus then its governing rotation is a space-time structure that builds on the non-linearity of k in the coupling parameter for the sine circle map. Therefore it is not a surprise that life is an oscillator if space-time is an oscillator. Further the Lorentz contraction or winding number can be equal to the golden mean governed by this equation[3],

5) $\alpha=\Omega=l/L=(L-1)/l=0.6180339$

The helix DNA is a double helical structure that uses the golden mean to spiral for one period of its double helix (21 angstroms/34 angstroms approximates the golden mean), so DNA can be thought of as a something that evolves due to non-linearity acting on space-time. Since the golden mean is the structural organization of DNA; life uses the golden mean for its structure and this is because of the equation number 5 that space-time can induce a golden mean structural change to matter when there is strong coupling K>1.

-

Experimental design and setup:

The Josephson Effect (proof of Alpha is equal to omega):

The Josephson ac effect is assumed to be a phase locking behavior of the damped periodically driven oscillations. Early experiments showed that when one increased the applied dc current, the current-voltage curve would develop a staircase structure with Shapiro steps[2]. The Josephson junction is two superconductors divided by a central strip where there is electron tunneling. A current generates energy across the junction when in conjunction with an applied microwave field. Fig 4 is the values.

system	periodically forced oscillator	Josephson junction
governing law	classical Newtonian mechanics	classical RCSJ effective circuit
θ	angle of oscillation	macroscopic quantum phase difference
ω_e	frequency of external force	frequency of microwave field
α	momentum of inertia	$\hbar C/2e$; C: capacitance
β	viscous damping	$\hbar/2eR$; R: resistance
$\gamma \sin\theta$	restoring torque	Josephson current
A	applied constant torque	applied dc current
$B \cos\omega_e t$	applied time-varying torque	applied microwave field

Fig. 5 Table of Values for Josephson junction in a microwave field.

By defining a Josephson frequency to be wj=dθ/dt and the energy difference across the junction may be written as

6) **E= hwj where h is Planks constant**

When integral Shapiro steps occur **wj=n we** and the last equation turns out to be

7) **E=nh we**

which implies that the energy difference for the tunneling of Cooper pairs is quantized in units of h we (the external microwave field), and the sub harmonic steps give

8) **E=(n/m)h we**

This behavior is compared to the steps found in a self-similar staircase where winding number is compared to the omega values and are found to be self-similar.

The winding number is the phase locking behavior in the circle map

9) **W=lim (θn-θo)/n**
   (n->inf)

The winding number physically measures the frequency ratio of the system and the external perturbation. Finally the energy across the junction is found to be

10) **E=hwj=h<dθ/dt>=h lim (2π(θn-θo)/nT)=Wh we**
    (n->inf)

The quantization relations can thus be interpreted by the phase locking behavior in which the frequency ratio W=n/m and the integral quantization when m=1, so that W=n=wj/we. This suggests that the nonlinear dissipative phase locking model provides a macroscopic fractional energy quantization interpretation for the super conducting Josephson ac effect under an applied microwave periodic field. The winding number of the map interprets the fractionally quantized energy levels by the frequency ratio of two nonlinearly coupled oscillations[2].

Analysis

Results of the the experiment.

The first experiment proves that alpha is indeed equal to omega as the energy difference defined by the steady state Josephson frequency is

12) E=hwj=Wh we

When $W=\Omega$ when the K=0 then we have $E\Omega=E'$ or $E'/\Omega=E$ where E' is the stationary energy of the quantum relationship but as we see this equation is the mass energy equation for Einstein ($E'/\alpha=E$). The omega values connect the frequency ratio of the two systems and therefore must be equal to alpha. So $\alpha=\Omega$ is proven true and is also the quantum number for the system. The Lorentz contraction is then quantized

13) hwe=W h we

thus we have given harmony to two major theories relativity and quantum mechanics as the Lorentz contraction means that this new theory of space-time can explain quantization.

As well we see the inverse of frequency is time and thus we=2π/T2 and wj=2π/T1 and therefore

14) T2/T1=W=Lorentz contraction=

So time dilation and energy have proved the ratio for omega is indeed equal to alpha a Lorentz contraction. Therefore our theory that space-time can be described by simultaneous rotation to create a straight geodesic line on the torus is proven true. Since the Lorentz contraction is a quantum number we can say that space-time is a harmonic oscillator that is quantized.

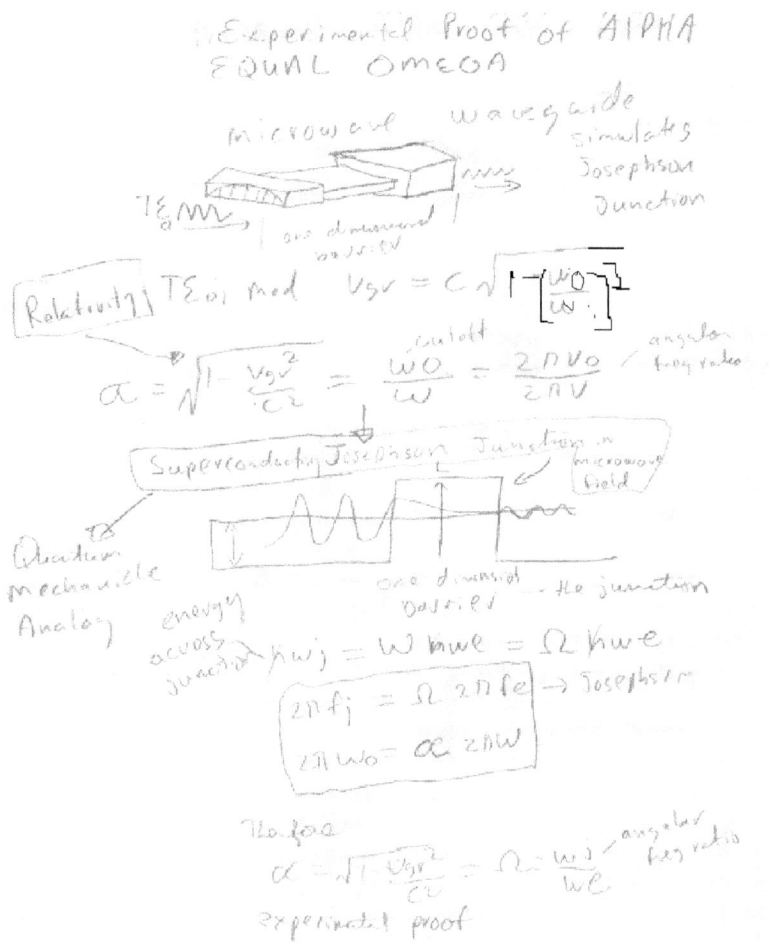

Fig 6) As we can see alpha orrega is derived from waveguide simulating the Josephson barrier

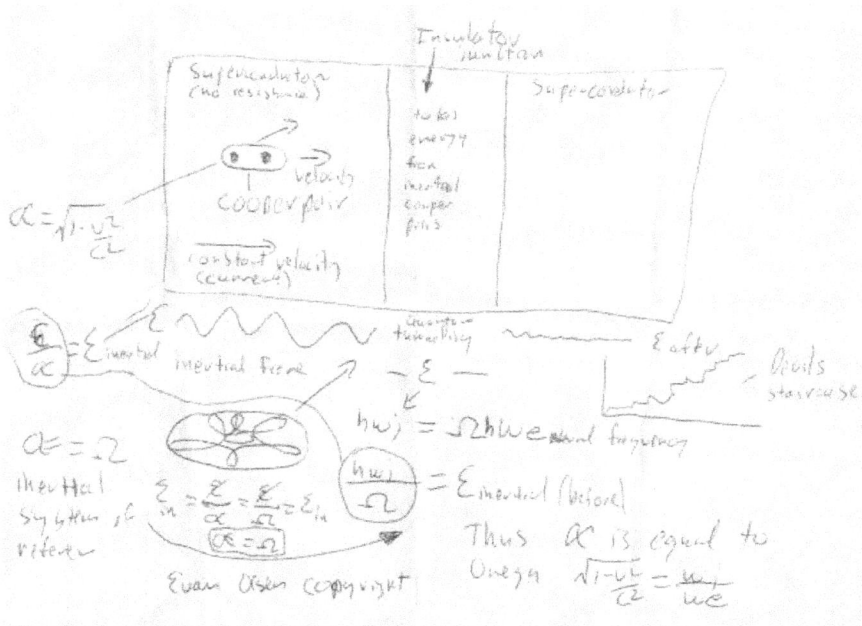

Fig 7) E/omega=E/alpha and thus the inertial motion proves alpha equal omega across the junction.

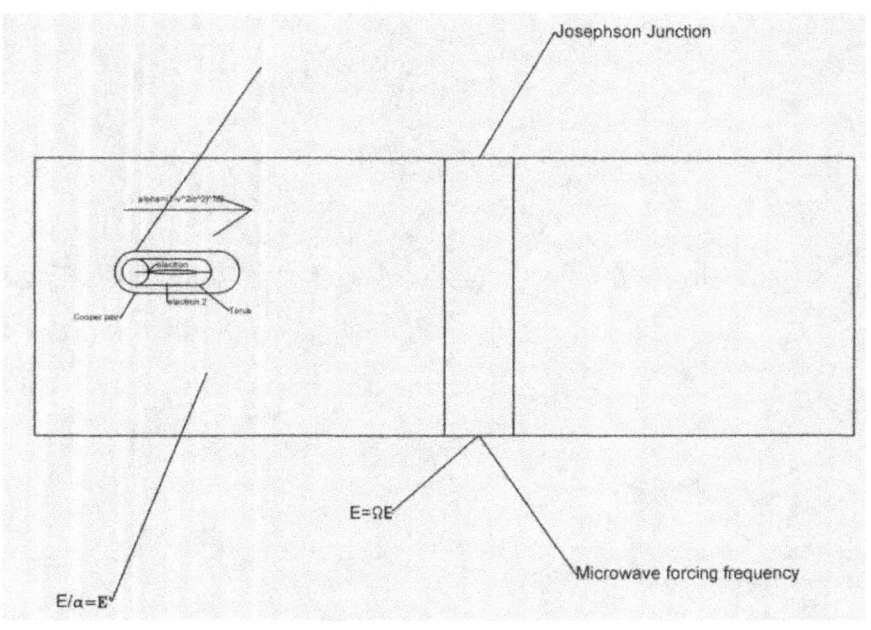

Fig 8 The E's cancel and thus alpha equal omega.

The autocad drawing of the superconducting Josephson junction in a microwave field (new theory of absolute zero). Thus the cooper pairs lose energy as they collide across the junction energy gap delta and this is true for the cooper pairs and thus E/alpha=E/omega and thus alpha equal omega and they give of current so there must be a collision. It important to note here that this is just the cooper pairs for an object that's on the wind on the torus alpha equal omega the quantized geodesic harmonic oscillator's of space-time give the object energy incrementally and instantaneously changing the velocity without violating energy conservation and not dependent on the speed of light.

-

Proof that this is faster then speed of light.

The next diagram focuses on an inelastic collision that is one dimensional and that happens on the torus that the electrons are colliding faster then light. Each oscillator sees the other at rest and the two motions are one combined motion.

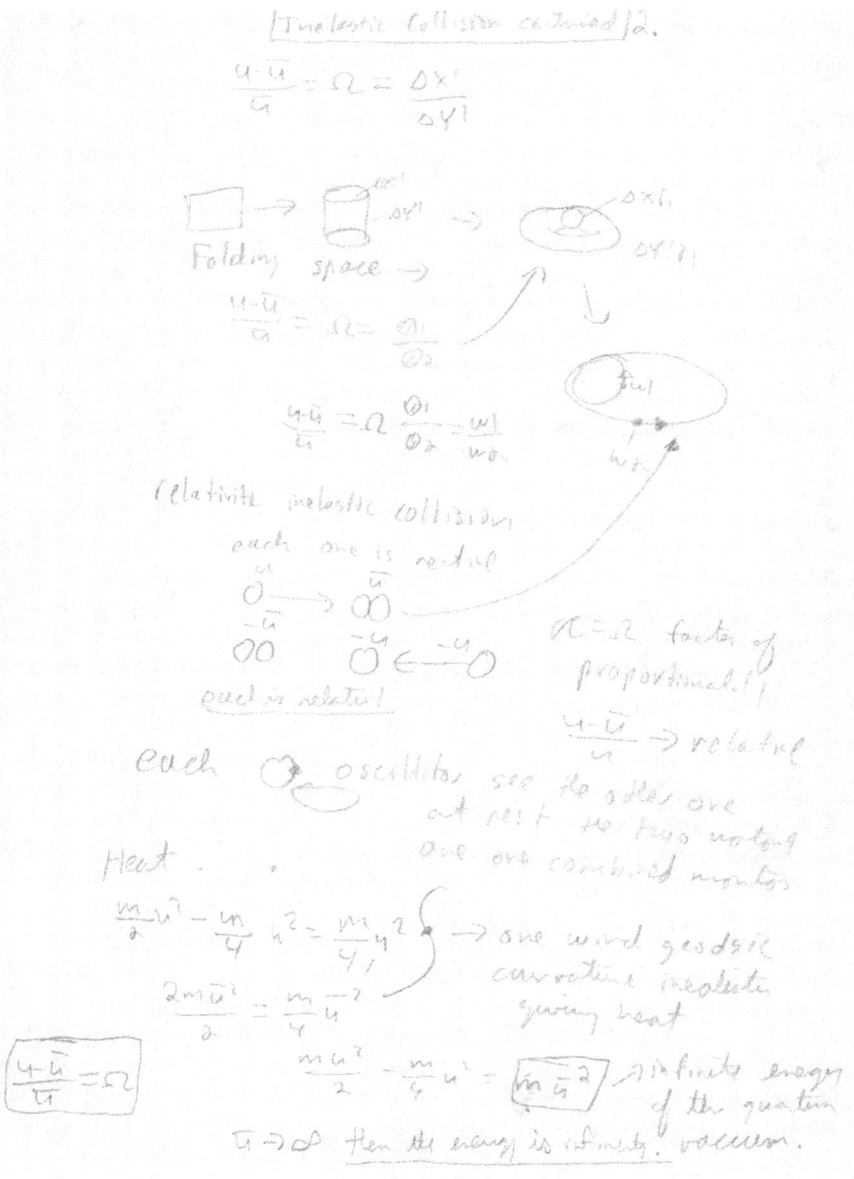

Fig 9) Proof that omega is an inelastic collision that two particles make on the torus that can go faster then light from Max Born equation found in "Einstiens theory of relativity".

Fig 10 proof that inelastic collision goes forwards
and backwards and that the velocities can be faster then light.

That in $\alpha=\Omega$ that omega can replace alpha and the inelastic collision equation for a Lorentz transform is $(u-u^1)/u^1=\Omega$ **as shown in Max Borns book) and thus the combined motion is a collision whose velocity that does not depend on the speed of light u is the velocity of the particles moving the entangled distance before the collision** and u' is the velocity of the combined particles. The proof here comes from the fact that the collision is relative to the system of observation it is possible to view the collision as two different systems with each particle at motion or at rest because the motion is relative. Thus this u value isnt set as a maximum of constant of light. Further entanglement can be thought of as two particles whose distance between them is a Lorentz transform because the distance is increasing between them at constant velocity.

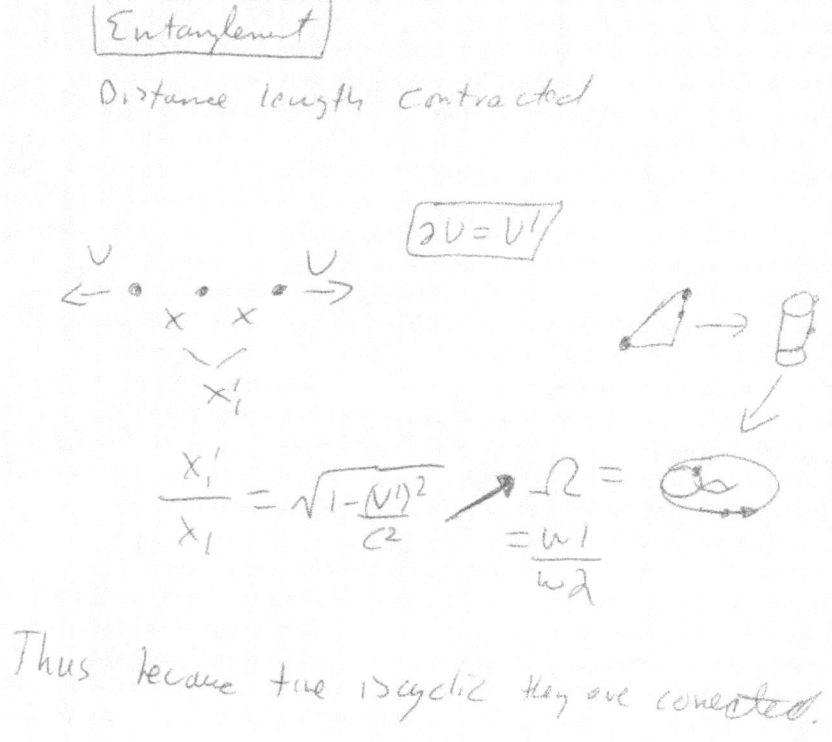

Fig 10) The solution to entanglement proving that
the wave state collapses simultaneously and thus uncertainty is chaos.

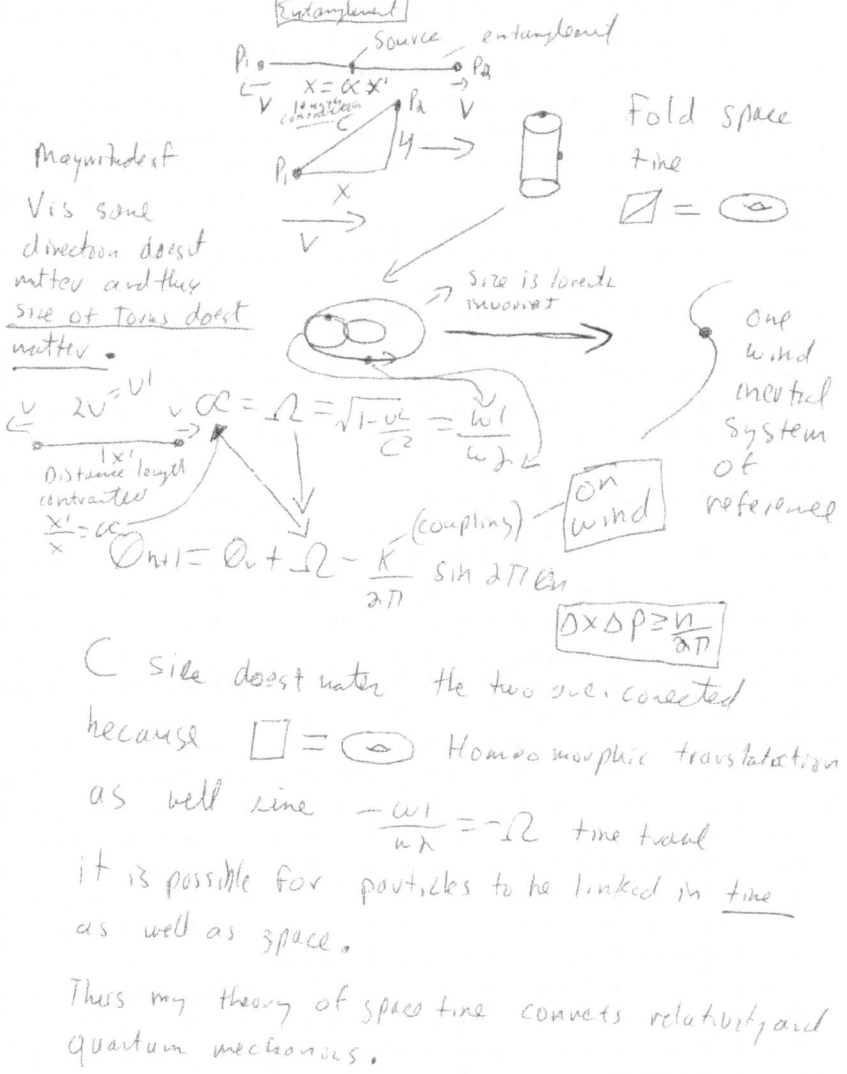

Magnitude of V is some direction doesn't matter and thus size of Torus doesn't matter.

$$\mathcal{C} = \Omega = \sqrt{1 - \frac{v^2}{c^2}} = \frac{\omega'}{\omega \lambda}$$

$2v = v'$

$x = \alpha x'$ length contraction

$\frac{x'}{x} = \alpha$

Distance length contracted

Fold space time

$\square = \bigcirc$

Size is lorentz invariant

One kind inertial system of reference

$$\Theta_{n+1} = \Theta_n + \Omega - \frac{K}{2\pi} \sin 2\pi \Theta_n$$ (coupling)

on wind

$$\Delta x \Delta p \geq \frac{n}{2\pi}$$

C size doesn't matter the two are connected because $\square = \bigcirc$ Homeomorphic translation as well since $-\frac{\omega'}{\omega\lambda} = -\Omega$ time travel it is possible for particles to be linked in time as well as space.

Thus my theory of space time connects relativity and quantum mechanics.

Fig 11) Further we see how the folding of space connects the two particles and this is an isomorphic translation.

Further through the help of "Very High absorption in Superconductors" by M.J Beckingham we see there is a third equation delta.

$$\boxed{\alpha = \Omega = \Delta}$$

due to
collisions

$\left(1-\frac{I}{T_c}\right)$ heat energy

$\left(\frac{h\omega}{KT}\right)^2$

linear

heat

$$\mathcal{E} = mc^2 = \frac{1}{2}m\left(\frac{u-\bar{u}}{\alpha}\right)^2 = \frac{1}{2}KT$$

$$\sqrt{1-\frac{I}{T_c}} = \left(\frac{h\omega}{KT}\right) = \sqrt{1-\frac{I}{T_c}} = \frac{h\omega}{KT} = \frac{u-\bar{u}}{u} = \Omega$$

$$= \Omega = \frac{h\omega}{n\hbar_A}$$

Thus $\alpha = \Omega = \Delta$ is related to heat energy of the collision of the electromagnetic quantum vacuums or spacetime

$$\frac{1}{2}mv^2 = \frac{1}{2}KT$$

$$\frac{mu^2}{X} = \frac{T}{T} = \frac{v_i^2}{v_{i2}}$$

$$\frac{mu^2}{X}$$

$$\underset{(\alpha)}{\sqrt{1-\left(\frac{v_i}{v_2}\right)^2}} = \underset{(\Delta)}{\sqrt{1-\frac{T}{T_c}}}$$

Thus spacetime is composed of quantized geodesic harmonic oscillators !!!! → that are entangled Corentz relativity

$$\alpha = \Omega = \Delta \rightarrow \text{every gap}$$
quantum   entangled particles
harmonic.

Fig 12) Alpha=Omega=Delta from relationship in M.J Beckingham

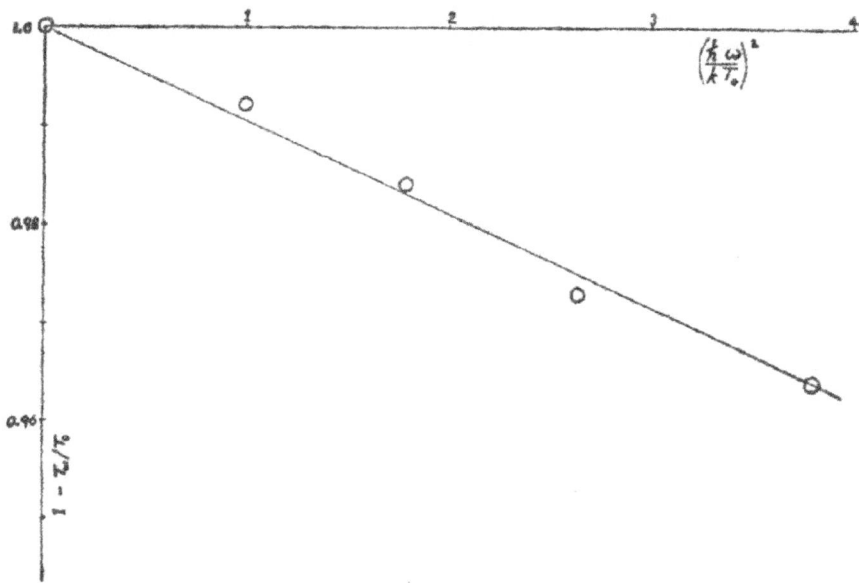

Fig Fig13 A diagram that explains energy gap relationship on temperature.

As we can see for these diagrams delta does not have to have the constant of light for its ratio for temperatures thus delta can be faster then light Further if one solves the relationship its $(1-t/to)^{1/2}=hw/KT0$ which is just a frequency ratio is $hw1/hw2$ of quantized geodesic harmonic oscillators. This is now a quantum tunneling equation that relates the alpha macro to the micro delta and this is true for superconductors and must also be true for the pi bonds in DNA that exhibit a delta relationship for their bonds if indeed a strange attractor is the breakup of harmonic omega motion to a strange attractor a factor of proportionality between order and chaos. Further it's obvious as the temperature gets close to T0 the energy on the graph increases to the right just like for a Lorentz transform. Further if $(u-u')/u'=\Omega(1-t1/t2)^{1/2}$ **where t2 has no upper limit which means u has no upper limit which also solves that velocity can be faster then light. u-u'/u' $=(1-u^2/c^2)^{1/2}$ and thus u is bounded by c but for omega.**

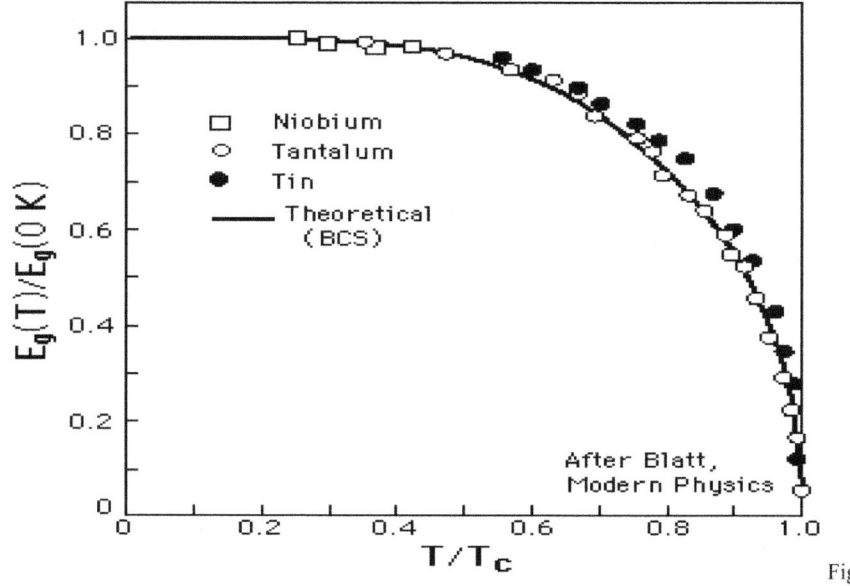

Fig

http://hyperphysics.phy-astr.gsu.edu/hbase/Solids/hitc.html

$(1-t1/t2)^{1/2}$ t2 has no limit t1 has no constant upper limit.

Therefore things can go faster then light and collide faster then light.

Conclusion:

The idea that space-time is an oscillator is paramount in the idea of $\alpha=\Omega=\Delta$. If the space-ct axis can be replaced by an oscillation or rotation then one can state that the whole concept of coordinate geometry of space-ct frames joined together is in fact the same as the two minor and major rotating systems of the winding number of the circle map. The wind is the connection of the two systems and is in fact the factor of proportionality alpha. If this is true then the winding number should produce different values depending on mass by the equation $E^1/\alpha=E$ or $E^1/\Omega=E$ for energy and time. The answer to this is yes it does, if one examines the Josephson junction in a microwave field.

This proves that omega or winding number fractionally changes energy, frequency and time. Thus space-time must be an oscillator and can also be quantized on the map of a torus. If this is true then configurations of space-time (matter) must be oscillators. The torus is finite and unbounded. Our own expanding universe is also finite and unbounded. Therefore the motion (omega) on the circle map or torus must represent a Lorentz transformation in space-time. So it is true the concept of Alpha equal Omega equal delta brings Harmony to three major physical theories, relativity, quantum theory and chaos theory.

Another example of a space-time configurations leading to chaoticle oscillators, is due to the evolution of biochemistry on the molecule DNA whose basis is an oscillator that's existence is due to the evolution of space-time interacting with the chaotic universe. Evolution is nothing more than the manifestation of chaos on the space-time continuum. This is why DNA is a strange attractor and this is why DNA is a double helical winding number that approximates the golden mean. DNA is a winding coupled helix that displays chaos. This completes Harmony theory because space-time and DNA are harmonic oscillators where the latter is Harmony by chaos. As well the new type of Lorenz contraction may give rise to quantum randomness through K which has been interpreted as probability for several decades. Finally if alpha equal omega then a Lorenz transformation can equal the golden mean and since matter is a space-time configuration then it is this Golden mean on the circle map that allows for the possibility of life, as we see the Golden Mean configuration from DNA to the structural anatomy of living organisms. Therefore it is space-time that is responsible for life. Life is flowing equilibrium harmony structure caused by coupling and chaos in space time.

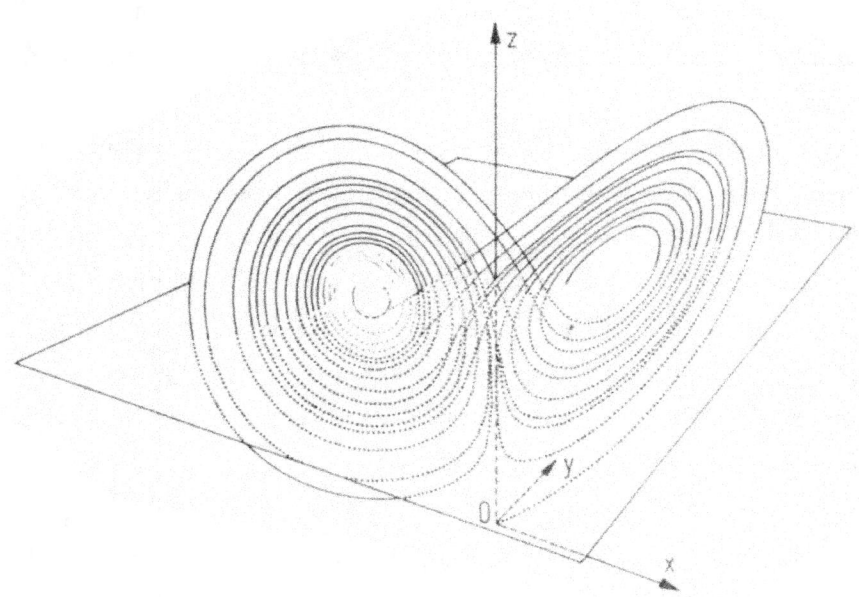

Fig 13 The strange attractor of lorenz.

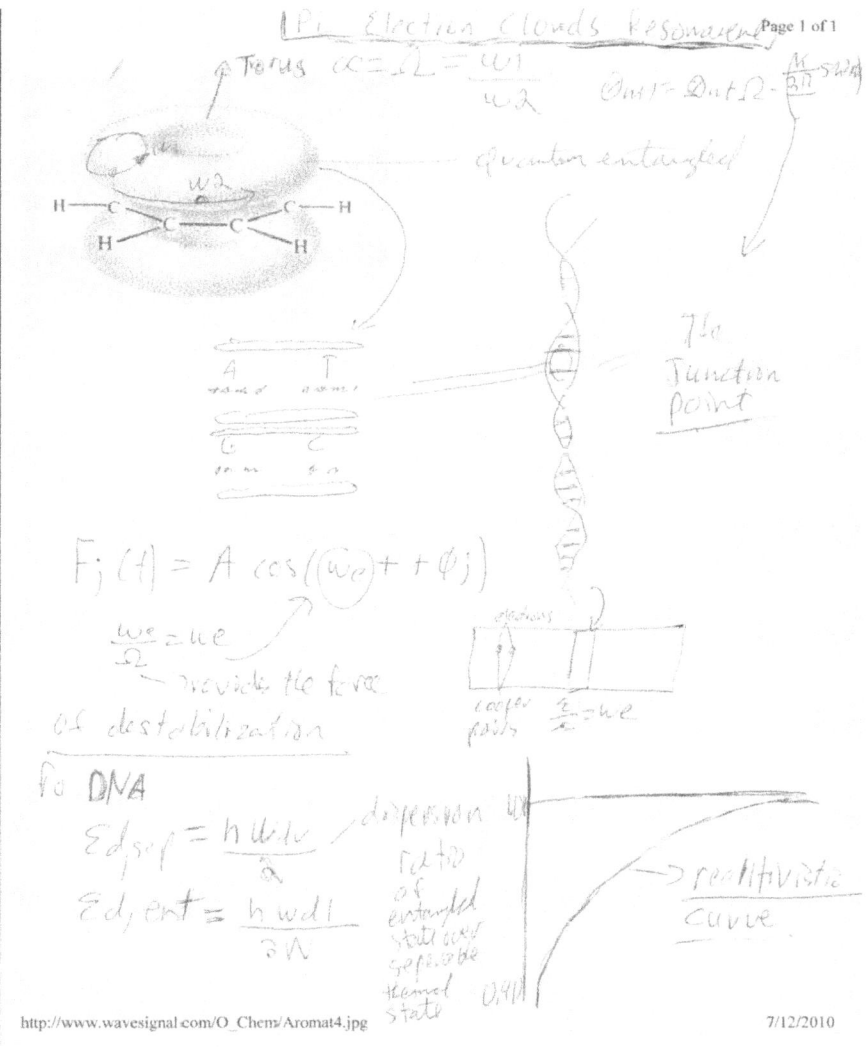

Fig 14 Elisabeth Rieper, Janet Anders and Vlatko Vedral "The Relevance of Continuos variable Entanglement in DNA" Center for Quantum Technologies National University of Singapore

Furthermore a Lorenz contraction changes matter a space-time configuration and omega of the circle map induces a Lorenz contraction then for mass-energy conversion you need to not only the bare winding number omega to produce a Lorenz contraction you need to change the structure of matter through the coupling parameter k. So you not only change energy and time but you change the structural organization of these through the coupling parameter k to form chaos. This is mass-energy conversion, we are not just changing the energy of the system but we are changing the way in which the energy is related or the Lorentz transformation alters the matter but the coupling parameter determines how the Lorentz contraction alters our matter. This is the fundamental theory that is stated in my book that space-time is the building block that acts through the non-linearity inherent in k and thus we have our different structural configurations of matter, through a Lorenz contraction and value k which determines the winding number, which is proven by the Josephson junction to be a Lorenz contraction. So the sine circle map equation is a Lorentz transformation that can be coupled to matter so as to change the configuration through perturbations inherent in energy exchange.

All this brings us to our new laws of space-time and DNA.

1. A Lorenz transformation for straight line motion is the same as the geodesic winding number on a circle map according to general relativity. This leads us to mass-energy conversion and time travel and the idea that velocity on the torus can be faster then light communication. In other words cooper pairs are colliding faster than the speed of light through the junction because the wind is not dependent on the speed of light. Thermal rectifiers in Josephson junctions have been shown to exhibit negative entropy while switching the magnetic field from clockwise to counter clockwise.

2. The winding number is in fact the representation of two oscillating systems meaning that space-time can represent a harmonic oscillator in four dimensional phase space. A Lorentz transfor-

mation is equal to the frequency ratio of circle map which is equal to the energy gap equation for superconductivity.

3. The winding number for the superconducting Josephson junction is the quantization. Thus in this special case the Lorenz contraction is the quantization and forms the inertial system reference before the junction turns into omega across the energy gap delta and proves spacetime is quantized.

4. If life and DNA is an oscillator and a strange attractor then it is the space-time torus that is the causation of life as a harmonic oscillator that bridges unto chaos due to increased coupling and the golden mean and the pi bonds exhibit quantum entanglement and this could cure cancer because we may test for the DNA we don't want and rip apart the dna with electromagnetic fields at the right junction point of the pi bonds.

5. It must be that $\Delta=(1-T1/T2)^{1/2}$ is the third equation as it is the determiner for the energy gap across Josephson junctions and is equal to an energy ratio (hw1/KTo) and shows a one to one correspondence to omega.

References:

[1]Born, Max "Einstein's Theory of Relativity". Dover Publications, Inc, New York p.170-308. 1965

[2]Wang T.C & Gou Y.S "A discussion on the Josephson Ac effect and the fractional energy quantization." Chinese Journal of Physics, April 1996 p.665-669

[3]Schuster, Georg " Deterministic Chaos: An introduction." Physik-Verlag GmbH, D-6940 Weinheim (Germany) (1984)

[4]"Very High absorption in Superconductors" by M.J Beckingham

[5]Elisabeth Rieper, Janet Anders and Vlatko Vedral "The relevance of continuos variable entanglement in DNA" Center for Quantum Technologies National University of Singapore

Figures Cited

Fig 1

Schuster Heinz Dr Deterministic Chaos. (Germany Physik – Verlag, 1984) p.119

Fig 2

Schuster Heinz Dr Deterministic Chaos. (Germany Physik – Verlag, 1984) p.116

Fig 3

hand drawn by Evan Olsen

Fig 4

Schuster Heinz Dr Deterministic Chaos. (Germany Physik – Verlag, 1984) p.92

Fig 5

Fig 6

hand drawn by Evan Olsen

Fig 7

Autocaded from Evan Olsen

Fig 8

Hand Drawn by Evan Olsen

Fig 9

Hand Drawn by Evan Olsen

Figure 10

Hand drawn by Evan Olsen

Fig 11

Hand drawn by Evan Olsen

Fig 12

Fig 13

Very High absorption in Superconductors" by M.J Beckingham

Fig 14

Schuster Heinz Dr Deterministic Chaos. (Germany Physik – Verlag, 1984) p.104

Chapter 2

Miscellaneous proofs and space-time continuum.

Algebraic
Proof for Alpha = Omega

$\alpha = \Omega$

$\theta > 45°$ so $\frac{Y}{X} > 1$ — Only true for these values

lets use

$\tan 50° = 1.192$

$\gamma = \frac{1}{\left(1 - \frac{v^2}{c^2}\right)^{\frac{1}{2}}}$

$1.192 = \frac{Y}{X} = \frac{1}{\gamma} \frac{\Delta Y'}{\Delta X'} = \frac{1}{\gamma} \tan \alpha'$

$1.192\, \gamma = \frac{\Delta Y'}{\Delta X'}$

Lets move to moving frame to torus
Translation

$r_1 < r_2$

radius of $\Delta X'$

radius of $\Delta Y'$

$0 < \Omega < 1$

$0 < \frac{\Delta X'}{\Delta Y'} < 1$

$\frac{1}{1.192} \gamma = \frac{\Delta X'}{\Delta Y'} = \frac{r_1}{r_2} \frac{\theta_1}{\theta_2}$

$\frac{1}{\gamma} = \Omega$

$.84\, \Omega = \frac{\Delta X'}{\Delta Y'} = \frac{r_1}{r_2} \frac{\theta_1}{\theta_2} = \frac{r_1}{r_2} \Omega$

$1 > .84 = \frac{r_1}{r_2}$ — aspect ratio is correct for ring torus radius

This is only true if $\Omega = \frac{\theta_1}{\theta_2}$

$\frac{1}{\gamma} = \alpha = \Omega = \frac{\theta_1}{\theta_2} = \sqrt{1 - \left(\frac{v^2}{c^2}\right)}$

aspect ratio
for ring torus radius

31

[Inelastic Collision continued]2.

$$\frac{u-\bar{u}}{u} = \Omega = \frac{\partial x^i}{\partial y^i}$$

Folding space →

$$\frac{u-\bar{u}}{u} = \Omega = \frac{\partial_1}{\partial_2}$$

$$\frac{u-\bar{u}}{u} = \Omega \frac{\partial_1}{\partial_2} = \frac{w_1^i}{w_2^i}$$

relativite inelastic collision

each one is rectel

$$O \xrightarrow{u} \quad \infty$$
$$\infty \quad \xleftarrow{\bar{u}} O \xleftarrow{-u} O$$

each is relatul

$$\Pi = \Omega \text{ factor of}$$
proportionality

$$\frac{u-\bar{u}}{u} \rightarrow \text{relatal}$$

each ⬯ oscillator see the other one
at rest so two notes
one one combined montos.

Heat

$$\frac{m}{2}u^2 - \frac{m}{4}u^2 = \frac{m}{4}y^2 \Big\}$$ → one word geodsk
curvature inelastic
giving heat

$$\frac{2m\bar{u}^2}{2} = \frac{m}{4}\bar{u}^2$$

$$\boxed{\frac{u-\bar{u}}{u} = \Omega}$$

$$\frac{mu^2}{2} - \frac{m}{4}u^2 = \boxed{m\,\bar{u}^2}$$ → infinite energy
of the quantum

$\bar{u} \to \infty$ then the energy is infinitly. vacuum.

32

The next diagram is

$$\boxed{\alpha = \Omega = \Delta}$$

$$\left(\frac{\hbar\omega}{kT}\right)^2$$

linear

due to
collisions

heat energy

heat

$$\varepsilon = mc^2 = \frac{1}{2}m\left(\frac{u-\bar{u}}{\alpha}\right)^2 = \frac{1}{2}kT$$

$$\sqrt{1-\frac{I}{T_c}} = \left(\frac{\hbar\omega}{kT}\right)^0 = \sqrt{1-\frac{I}{T_c}} = \frac{\hbar\omega}{kT} = \frac{u-\bar{u}}{u} = \Omega$$

$$= \Omega = \frac{\hbar\omega}{n\hbar\omega}$$

Thus $\alpha = \Omega = \Delta$ is related to heat energy of the collision of the electromagnetic quantum vacuums or spacetime

$$\frac{1}{2}mv^2 = \frac{1}{2}kT$$

$$\frac{mu^2}{x} = \frac{T}{T} = \frac{v_i^2}{v_{i2}}$$

$$\frac{mv^2}{x}$$

$$(\alpha) \qquad (\Delta)$$

$$\sqrt{1-\left(\frac{v_i}{v_2}\right)^2} = \sqrt{1-\frac{T}{T_c}}$$

Thus spacetime is composed of quantized geodesic harmonic oscillators !!!! → that are entangled
Corects relativity

$$\alpha' = \Omega = \Delta \rightarrow \text{energy gap}$$
quantum
harmonic.   entangled particles

Shows things are quantum entangled and the energy gap equation which is faster then speed of light for quantum tunneling is shown to be a linear approximation now the graph on the upper right side is generated by superconductivity in tin an article written in 1957. The question arises here as to why Einstein didn't see this relationship and I suspect that he didn't have the energy gap equation and he didn't have chaos theory to be able to combine the theories into a complete picture. Einstein went to his death bed trying to solve the grand unified equation. s

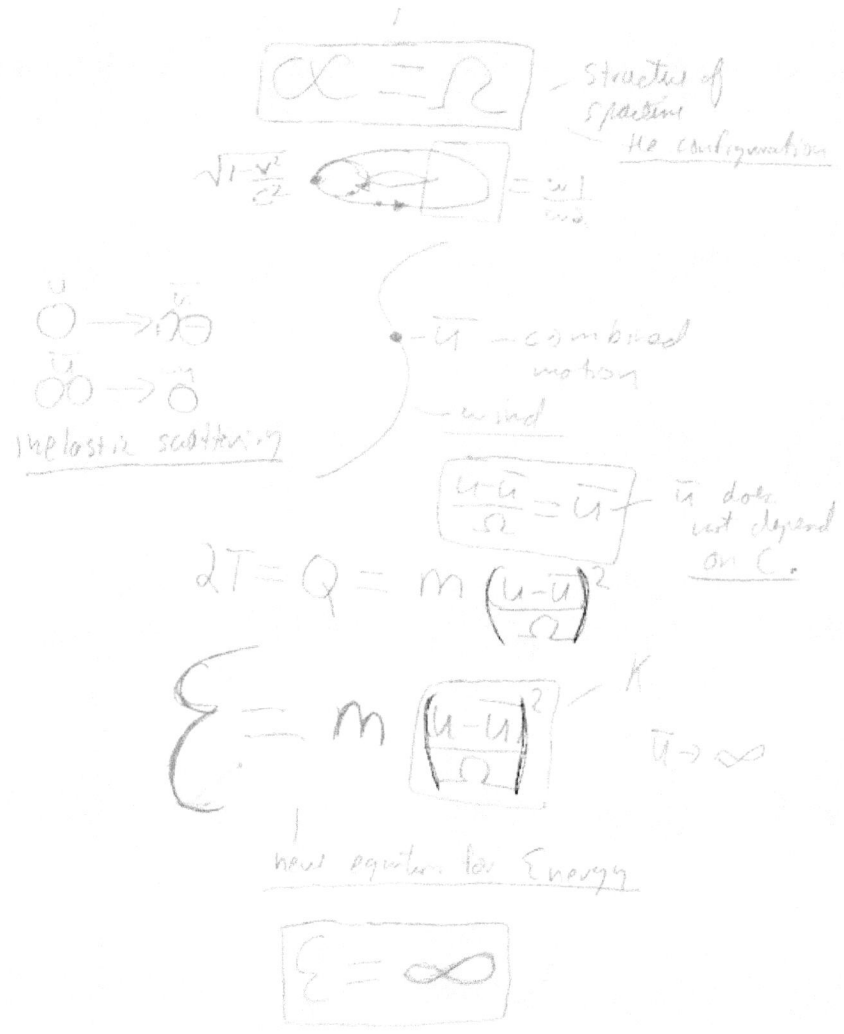

This picture here shows the constant can be bypassed and the equation for energy e=mc^2 can be rewritten as to show that the constant is now a new quantity with the C squared term being replaced with the u prime term and now classical physics gets a triumph is the kinetic energy department and if its electromagnetic in origin with the electrons with precise positions on the torus and they can generate chaos and show perfect symmetry when in combination for the curvature of the curve. The proof that the curvature of the curve is formed by a geodesic is shown on this diagram right after we see the quantized field equation for the metric tensor solved for in general relativity.

Quantized Field Equation

$$\sqrt{1-\Delta^2} = \frac{\ell q}{c^2}$$

$$\varphi = \frac{\lambda \varphi}{c^2}$$

$$\sqrt{1-\Delta^2} = \frac{\ell Gm}{r^2 c^2}$$

line element

$$ds^2 = -(\Delta^2) dt^2 + (\Delta^2) dr^2 + r^2 d\Omega^2$$

$$\Delta = \sqrt{1-\frac{2\ell Gm}{r^2 c^2}}$$

Schwarchild metric

$$\Delta = \sqrt{1-\frac{2Gm}{r c^2}}$$

Einstein field equation

$$G_{44} = R_{44} - \frac{1}{\lambda} R_{944}$$

$$g_{44} = 1 - \frac{2Gm}{r c^2}$$

$$G_{44} = R_{44} - \frac{1}{\lambda} R \Delta^{2\#}$$

$$\alpha = \Omega = \Delta$$

entangled cooper pairs electromaynetic

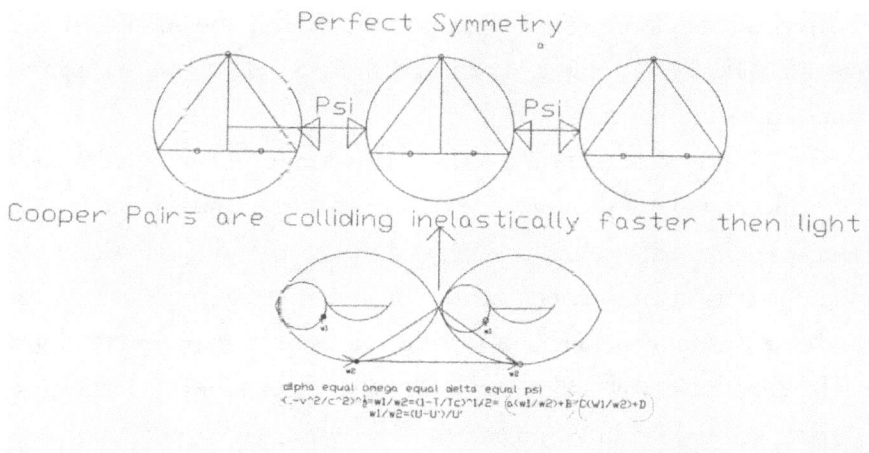

Perfect Symmetry

Psi   Psi

Cooper Pairs are colliding inelastically faster then light

alpha equal omega equal delta equal psi
<.-v^2/c^2>^½=w1/w2=(1-T/Tc)^1/2= [o(w1/w2)+B^C(w1/w2)+D]
w1/w2=(U-U')/U'

As one can see the connection between two torus's or two systems of reference is the connector the Mobius transform I label as psi. It is clear from this picture the relationship is real.

The formation of the equditriall
Cuse for tangent space

I somorphic Translation

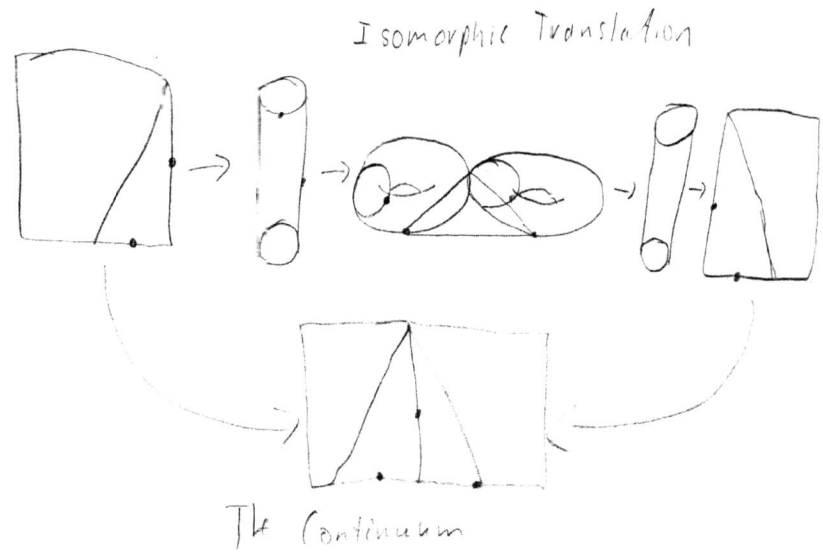

The Continuum

diagram depicting the calibration curve and the metric of the triangle is below with the two torus's coming together and the formation of the calibration curve.

The above picture shows that the two quantum entangled torus's are colliding together to form the curvature of the curve which is represented in tangent space as the N^j as the principal normal to the curve C and is obviously a generalization for a curve in a Euclidean space. Thus the Einstein's field equation is the representation of the curvature of the curve given by the metric tensor that determines the curvature of the space-time continuum and this curvature is now a quantum entangled configuration.

In space the reason that a force pushes an object that travels at constant velocity is that there is no resistance to flow by particles in void space just like a superconductor on earth the impedance to flow is caused by the random inertia of particles given by increasing temperature which means that temperature and heat and electromagnetism and gravity are one force if one thinks of space time as a superconductor.

Furthermore if these particles are colliding and there inertia can be faster then light not dependent on the constant then the propagation of signals between them is also faster then light because the calibration curve or curvature is not dependent on C.

Furthermore the diagonal of the parallelogram defined by the metric tensor which I have shown to be the frequency ratio of the torus is the X+Y vectors form a triangle on the calibration curvethat forms the metric of the space time continuum the metric of the type shown as g44

Calibration Curve

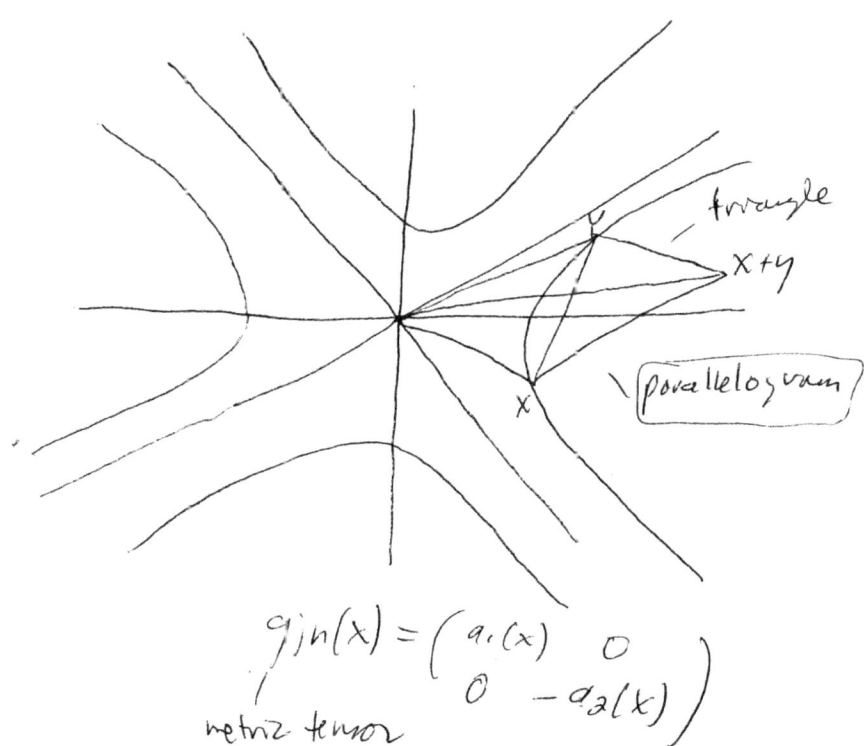

triangle

$x+y$

parallelogram

x

$$g_{jn}(x) = \begin{pmatrix} a_1(x) & 0 \\ 0 & -a_2(x) \end{pmatrix}$$

metric tensor

page 247 of Tensors, Differential
forms and Variational Principles

Entanglement

Distance length contracted

$$2U = U'$$

$\overset{U}{\longleftarrow} \bullet \quad \bullet \quad \bullet \overset{U}{\longrightarrow}$

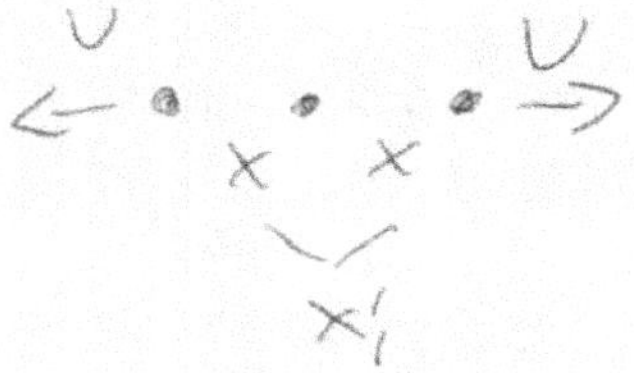

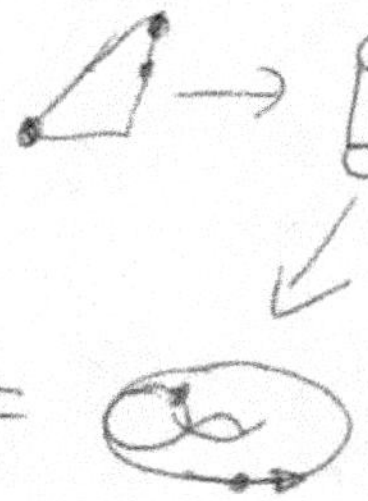

$x \quad x$

x_i'

$$\frac{x_i'}{x_i} = \sqrt{1 - \frac{(v')^2}{c^2}} \quad \nearrow \quad \Omega = \bigcirc$$

$$= \frac{w_1}{w_2}$$

Thus because they is cyclic they are connected.

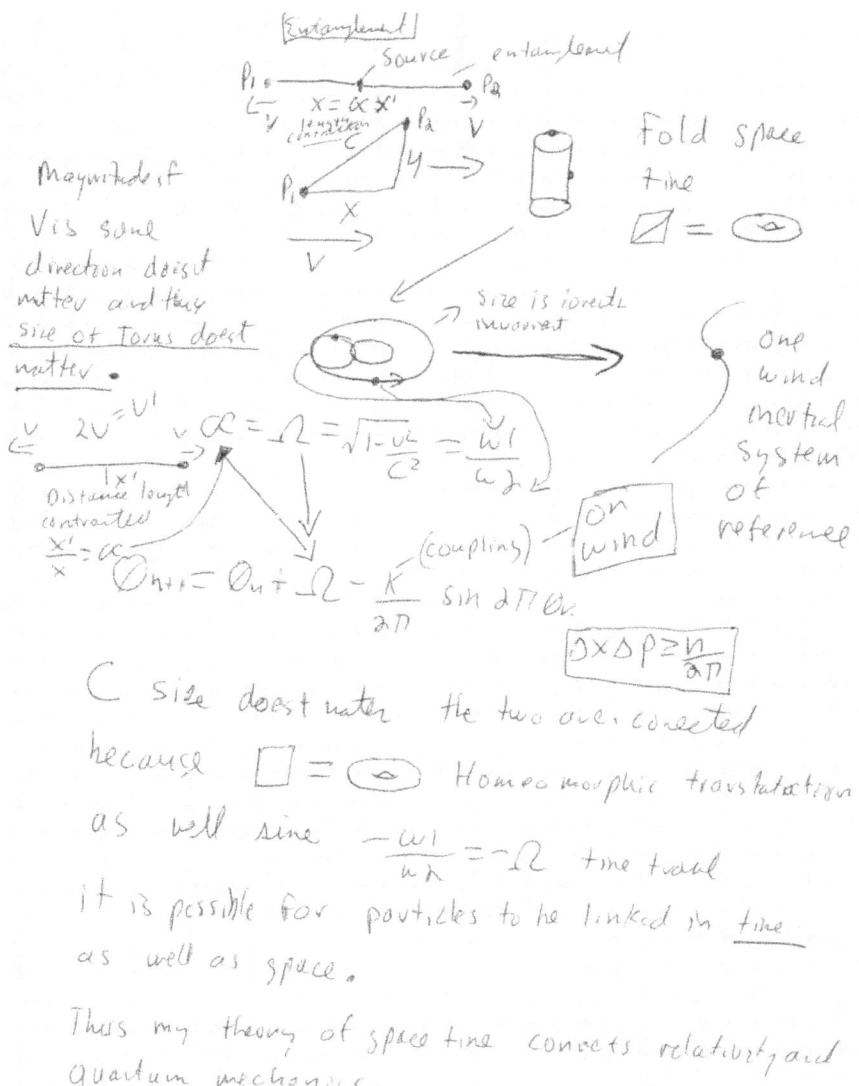

Entanglement

Source   entanglement

$P_1 \bullet \underline{\hspace{2cm}} \bullet P_2$

$x = \alpha \, x'$
length
contraction

P_2

$y \rightarrow$

Fold space
time

$\square = \bigcirc$

P_1

x

$V \rightarrow$

Magnitude of

V is same
direction doesn't
matter and thus
size of Torus doesn't
matter.

size is indeed
invariant

one
wind
inertial
System
of
reference

$2v = v'$

$\alpha = \Omega = \sqrt{1 - \frac{v^2}{c^2}} = \frac{\omega_1}{\omega_2}$

Distance length
contracted

$\frac{x'}{x} = \alpha$

(coupling)

On
wind

$O_{n+1} = O_n + \Omega - \frac{K}{2\pi} \sin 2\pi O_n$

$$\exists x \Delta p \geq \frac{n}{2\pi}$$

C size doesn't matter the two are connected

because $\square = \bigcirc$ Homeomorphic transformation

as well sine $-\frac{\omega_1}{\omega_2} = -\Omega$ time travel

it is possible for particles to be linked in time
as well as space.

Thus my theory of space time connects relativity and
quantum mechanics.

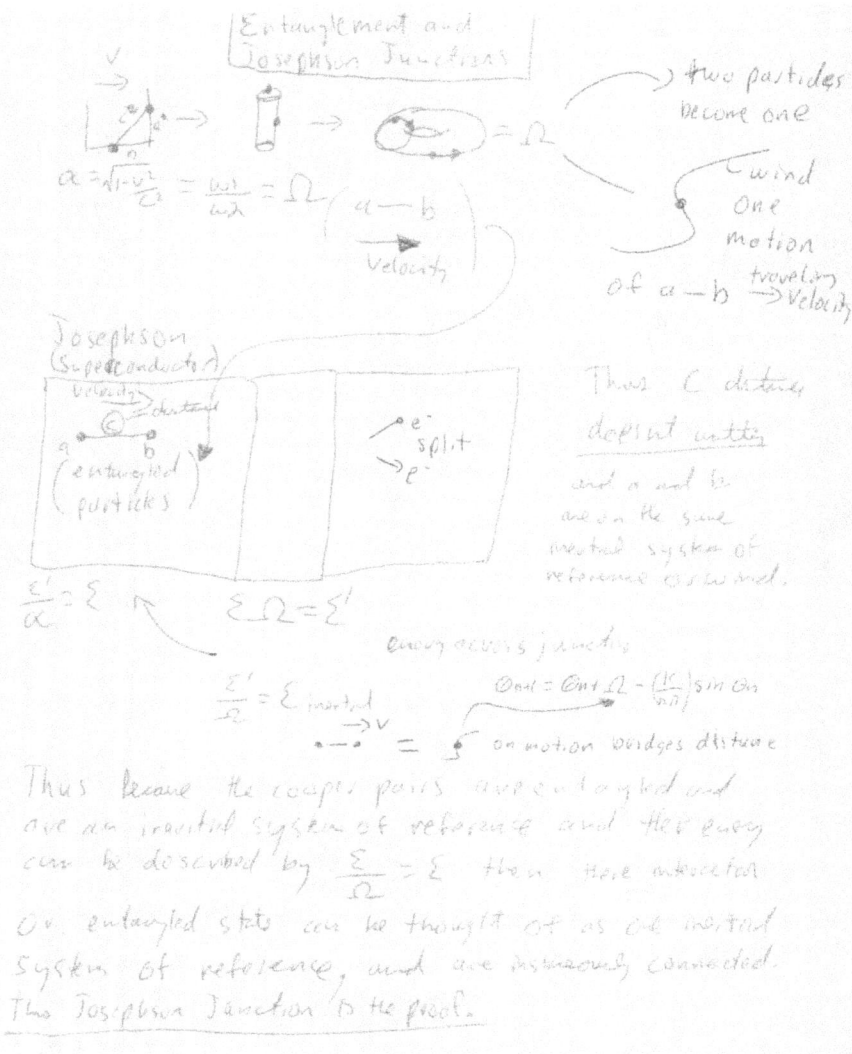

Is the proof that the nonlocal phenomenon of quantum entanglement forms a space time phenomenon which is composed of the metric tensor and forms a calibration curve for space time meaning that the theory of Einstein's relativity is not invalid but simply part of the first step and its possible to find a different configuration that bypasses the constant and forms the basis of space time and forms perfect symmetry.

The vectors that form the calibration curve need not be dependent on the speed of light the tensor calculus that forms the basis for this type of math need not have the speed of light subbed into it forms a separate proof altogether. This is where one has to ask is whether the either theory of relativity was right and the answer to this question is probably yes it's just the form of the either had not been formally defined no one knew about the effects of superconductivity and its relation to theory of relativity which are now abundantly clear to me is the proof of this theory and the reason that I have gotten six books all sequels to this theory of relativity. Einstein was wrong speed of light and the E=Mc^2 equation is not universal and thus the reason that he did not win noble for this reason is obvious he won it for the photoelectric effect. That light is a particle and these particles can be quantum entangled as for the Alan Aspect experiment done in the eighties. Further in his experiment the two particles that have spin interact simultaneously collapsing the wave function and thus bell's hidden variable is wrong the hidden variable is the folding of space time by an isomorphic projection square to torus preserving the moving frame of reference and constant velocity as a double harmonic oscillation. Further alpha equal omega equal delta is proof that quantum tunneling is a form of teleportation that can happen faster then light thus a person on the wind or omega can tunnel. Further we have related the micro delta to the macro alpha and thus large astronomical objects can tunnel faster then light. This is key we get a new technology that can tunnel objects that can travel forward or backwards in time.

Chapter 3

Time Travel, Antigravity

Of course going faster then light means that time travel is not only possible but is consistent with the theory. If one replaces the alpha with the omega then the negative time dilations can be put into the theory because omega is not under the square root. Furthermore this gives them negative mass and this proves that aliens not only possess the ability to move from present to past but also that there antigravity technology is dependent on negative mass particles giving them the negative force that pushes the crafts upwards in a gravitational field as shown

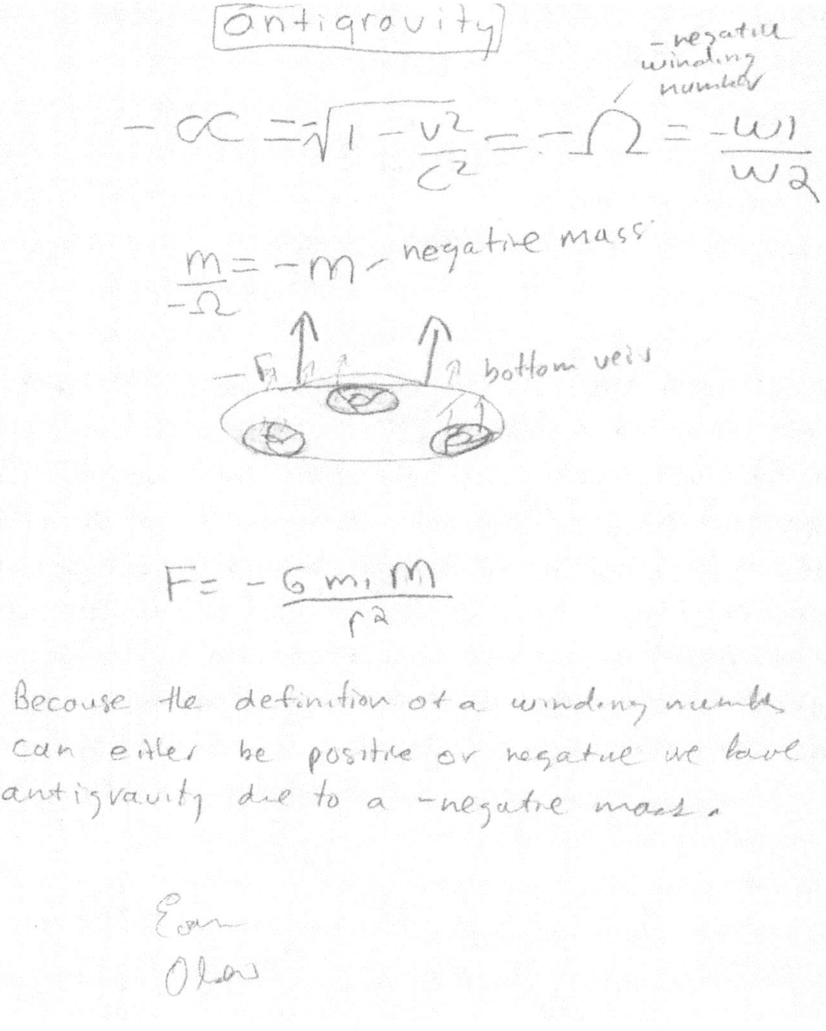

$$\boxed{\text{antigravity}}$$

$$-\propto = \sqrt{1 - \frac{v^2}{c^2}} = -\Omega = \frac{-\omega_1}{\omega_2}$$

– negative winding number

$$\frac{m}{-\Omega} = -m \text{ — negative mass}$$

bottom veiw

$$F = -\frac{G m_1 M}{r^2}$$

Because the definition of a winding number can either be positive or negative we have antigravity due to a –negative mass.

Evan
Olsen

Fig 1. Antigravity.

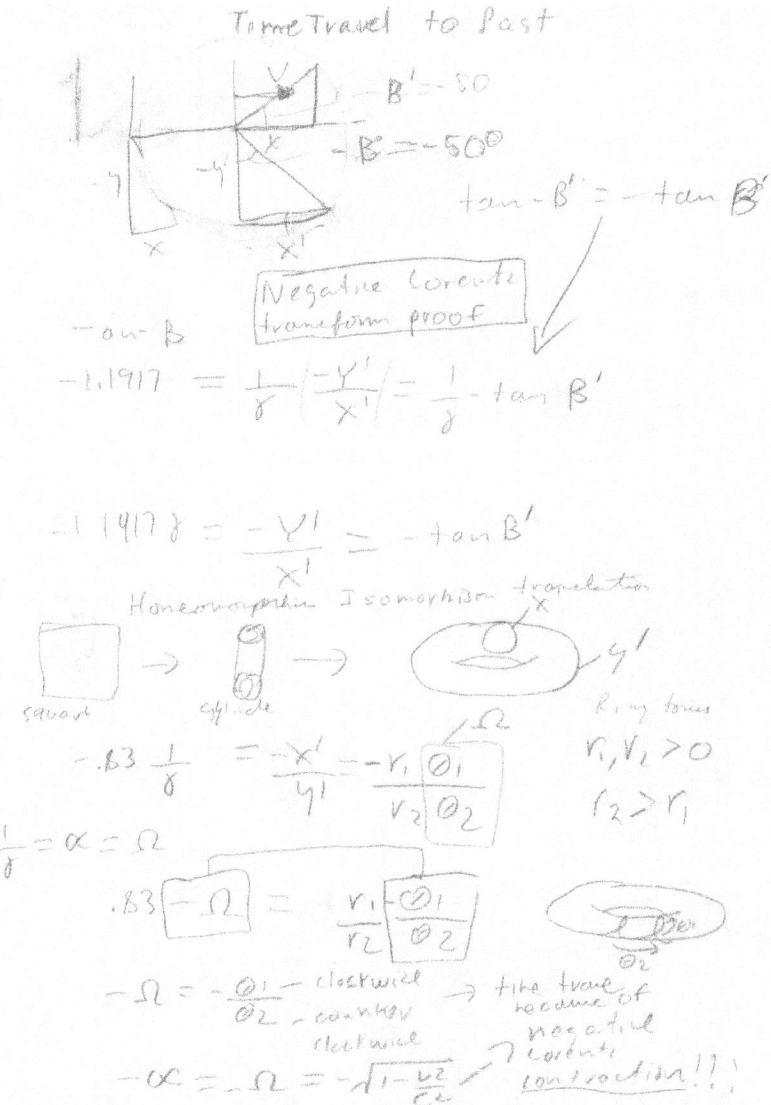

Figure 2. Time travel algebraic.

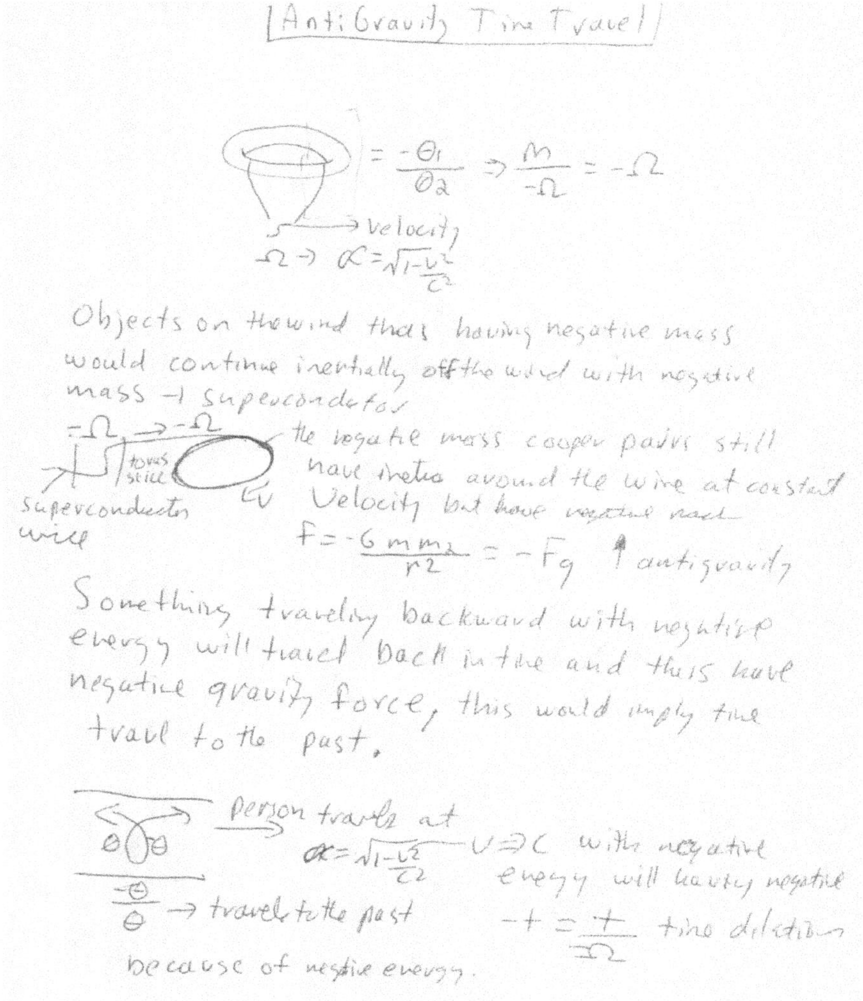

Figure 3. Antigravity time travel via a superconductor.

And of course ultimate proof for the whole thing comes from the Josephson junction that has a negative devils staircase shown in this diagram below.

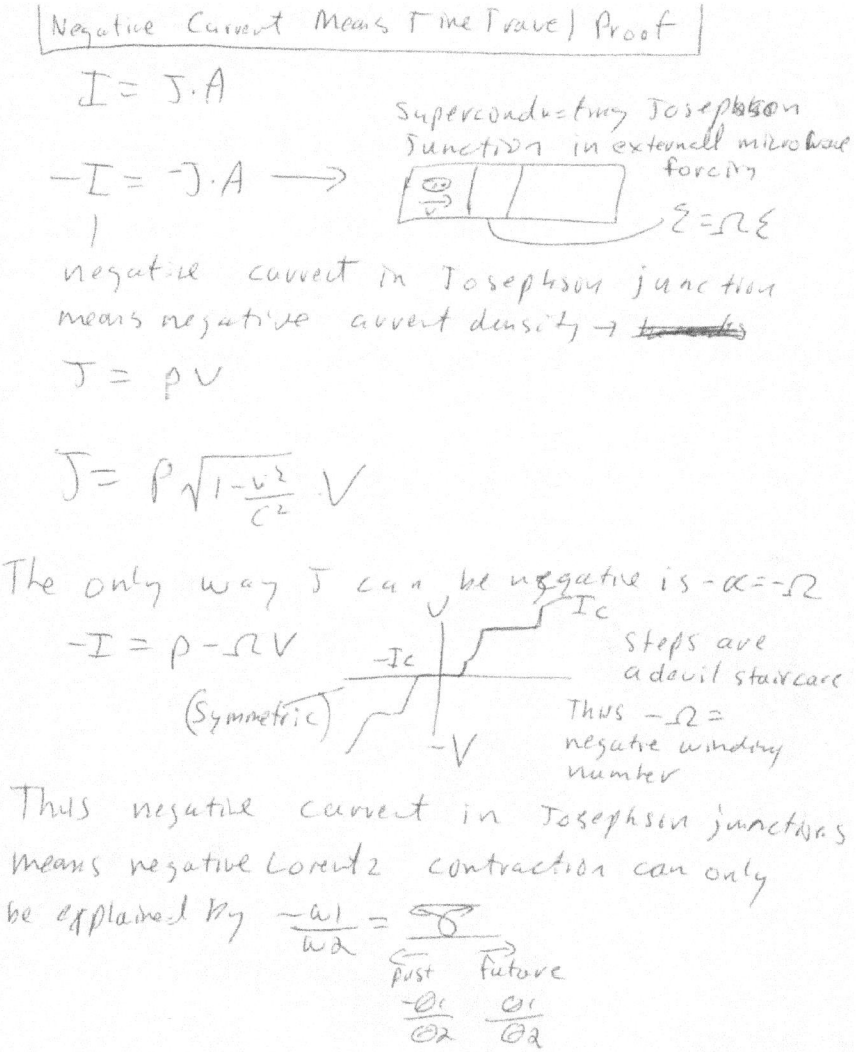

Negative Current Means Time Travel Proof

$$I = J \cdot A$$

$$-I = -J \cdot A \longrightarrow$$

Superconducting Josephson Junction in external microwave forcing

$$\xi = \Omega \xi$$

negative current in Josephson junction means negative current density → ~~things~~

$$J = \rho V$$

$$J = \rho \sqrt{1 - \frac{v^2}{c^2}} \, V$$

The only way J can be negative is $-\alpha = -\Omega$

$$-I = \rho - \Omega V$$

(Symmetric)

I_c

$-I_c$

$-V$

Steps are a devil staircase

Thus $-\Omega =$ negative winding number

Thus negative current in Josephson junctions means negative Lorentz contraction can only be explained by $\frac{-\omega_1}{\omega_2} = \frac{\Theta}{\Theta}$

Past   Future

$\frac{-\Theta_1}{\Theta_2}$ $\frac{\Theta_1}{\Theta_2}$

Figure 4. The current means that cooper pairs are flowing from future to past in the junction the only way to explain this is if its relativized is if the structure of the cooper pair is indeed the circle map or torus and thus can have negative winds to the past

49

If time were going backwards then the flow of heat must be reversed this has been proven in Josephson junctions in squids that have a magnetic field passing through them and thus heat flow has reversed breaking the second law of thermodynamics. The proof for this comes the article

Efficient phase-tunable Josephson thermal rectifier

M. J. Mart__nez-P_erez1, a) and F. Giazotto1, b)NEST, Istituto Nanoscienze-CNR and Scuola Normale Superiore, I-56127 Pisa, Italy

The cooper pairs are colliding in elastically to form a straight geodesic curve or line and this line is the inertial system of reference and this can be negative meaning that time is indeed cyclic. In the article the reversing of the magnetic field through the junction changes the flow of heat as predicted by my time travel equation on the wind. A negative wind would have negative heat flow. As seen below is a thermal rectifier that was predicted to reverse time flow and it does indeed do that.

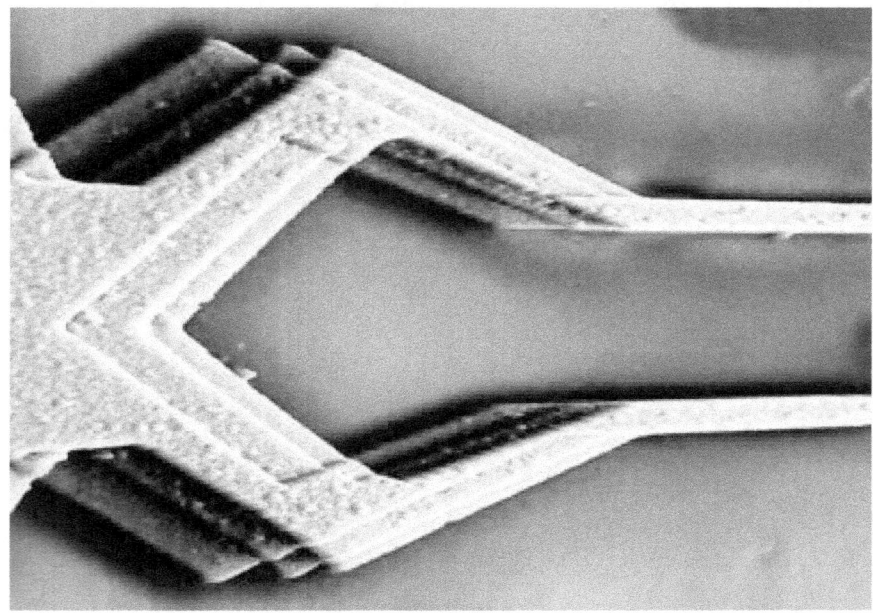

http://physicsworld.com/alheat-flows-across-josephson-junction

Efficient phase-tunable Josephson thermal rectifier

M. J. Mart__nez-P_erez1, a) and F. Giazotto1, b)NEST, Istituto Nanoscienze-CNR and Scuola Normale Superiore, I-56127 Pisa, Italy. Diagram of thermal rectifier.

Negative-Mass Hydrodynamics in a Spin-Orbit–Coupled Bose-Einstein Condensate M. A. Khamehchi,1 Khalid Hossain,1 M. E. Mossman,1 Yongping Zhang,2,3,* Th. Busch,2,† Michael McNeil Forbes,1,4,‡ and P. Engels1 Proves that negative mass is real another prediction I have shown that is proven true. Negative mass proves that time travel is real because the only way to produce it is to change the winding number to negative which is the solution to quantum entanglement and changes with spin change which is what this experiment has shown. One final note is that superconducting radio frequencies through delta replaced with omega could allow us to send radio frequencies to the past and receive from the future. Perhaps its possible to modelock through time as well a space.

-

Reversing the thermodynamic arrow of time using quantum correlations Kaonan Micadei, John P. S. Peterson, Alexandre M. Souza, Roberto S. Sarthou, Ivan S.Oliveira, Gabriel T. Landi, Tiago B. Batalhão, Roberto M. Serra and Eric Lutz

"By revealing the fundamental influence of initial quantum correlations on time's arrow, our experiment highlights the subtle interplay of quantum mechanics, thermodynamics and information theory. It further emphasizes the limitations of the standard local formulation of the second law for initially correlated systems and offers at the same time a novel mechanism to control heat on the microscale. It additionally establishes that the arrow of time is not an absolute but a relative concept that depends on the choice of initial conditions. While we have observed reversal of the arrow for the case of two spins, numerical simulations show that re-versals may also occur for a spin interacting with larger spin environments [25]. Thus, an anomalous heat current does not seem to be restricted to extremely microscopic systems. The precise scaling of this effect with the system size is an interesting subject for future experimental and theoretical investigations. Our results on the thermodynamic arrow of time might also have stimulating consequences on the cosmological arrow of time [30]." Which is in agreement with my arrow of time and negative omega. $\alpha=\Omega=\Delta$ since alpha is a macroscopic its possible to send large objects back in time.

-

Chapter 4

The Spaceship

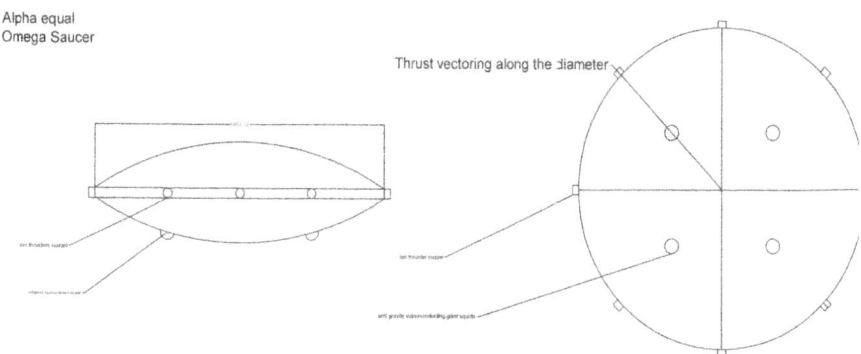

Alpha equal
Omega Saucer

Thrust vectoring along the diameter

Fig 1 The overall spaceship.

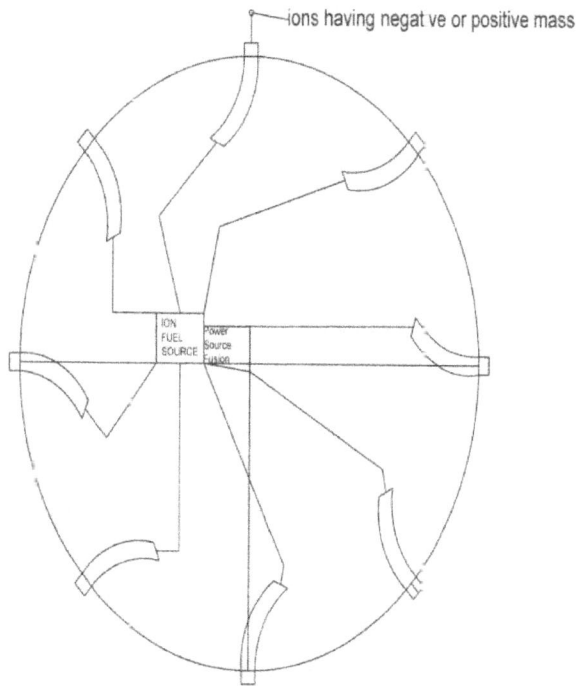

ions having negat ve or positive mass

ION
FUEL
SOURCE

Power
Source
Fusion

Fig. 2 Thruster Layout.

As we can see the cooper pairs travel towards the junction with constant velocity thus being an inertial system of reference and the energy is a fraction after the junction by E thus equating the two terms of alpha and omega.

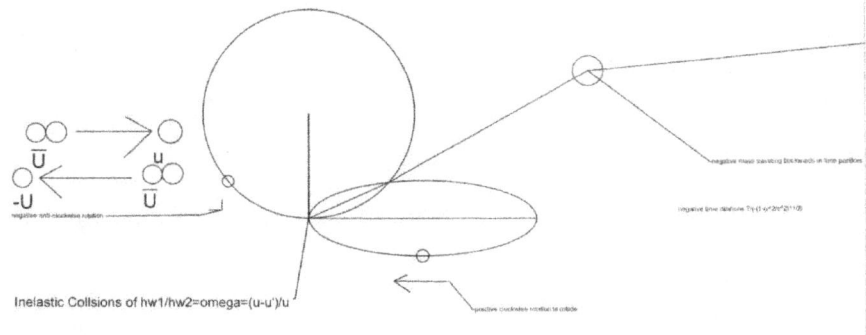

Fig 3 Diagram of omega inelastic collision on the torus.

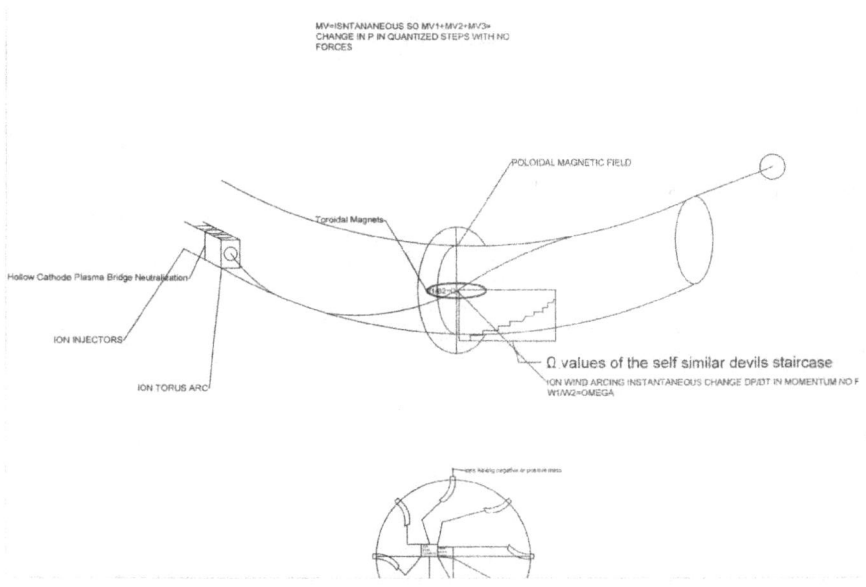

Figure 4. Thruster mechanism for increasing relativistic speeds.

These spacecraft can travel at relativistic speeds and are capable of shooting out negative mass particles that can put cause the spacecraft to travel backwards in time.

The important equation here is $M(v+vx)=Mo/((1-(v+vx)^2/c^2)$ the change in momentum is a change in the constant alpha and therefore omega and the above equation does not need a force. It possible to change the system of reference means were changing the momentum of the collision. In no time because the distance for entanglement divided by time is the velocity and the velocity after that collision is the particle stuck together which I prove is fusion but for this changing the ratio of omega changes the velocities and the collision $(u-u')/u'=omega$. Its possible to change the inertial system of reference by just changing alpha which changes the collision of the particle's on the wind with no mention of the force just the mass and energies and velocities.

Matter waves and faster then light.

From the louis debrolie equation we can get $(u-u')/u')h/(m)=u$ where

$$((u-u')/u')=omega=w1/w2$$

or

$$(u-u'/u)=delta=(1-t1/t2)^{1/2}$$

where there is no upper limit so the velocity can be faster then light for a matter wave.

–

Chapter 5

Multiple Dimensions and teleportation between Space-time inertial frames.

If things are generated from fractal's it means there are replications in the paths geometry that are angled from a common point which means there are different mathematical constructions of some geometry that are replicated around a mathematical equation that changes space-time according to external angles that connect these fractal connected dimensions. These external angles are equivalent to circle maps and thus a Mobius transforms which connects to sides of a strip meaning there is a door to other dimensions can be proven to be fractal possible and thus a theory of time travel is thus in agreement with a multiple timeline universe hypothesis the proof that devils staircases are fractal is also proof of this hypothesis. The proof that Mobius transforms connect circle to circles and triangles to triangles means that the transforms that form the continuum between them connect triangles like a fractal connecting triangles on a multiply connected fractal diagram. Spinors which describe the spin of electrons is changed by Mobius transform.

Spinor is a quantity resembling a vector or tensor that is used in physics to represent the spins of fermions. I just had a major breakthrough the solution to the Dirac equation in curved space-time is a spinor and the Mobius transforms are spinors and homeomorphisms they connect the omega cooper pairs and this has been solved in this chapter. Now I have the physical mechanism for the Mobius transformation. The next diagrams put in the back of my previous book are the proof that fractals are responsible for most viruses.

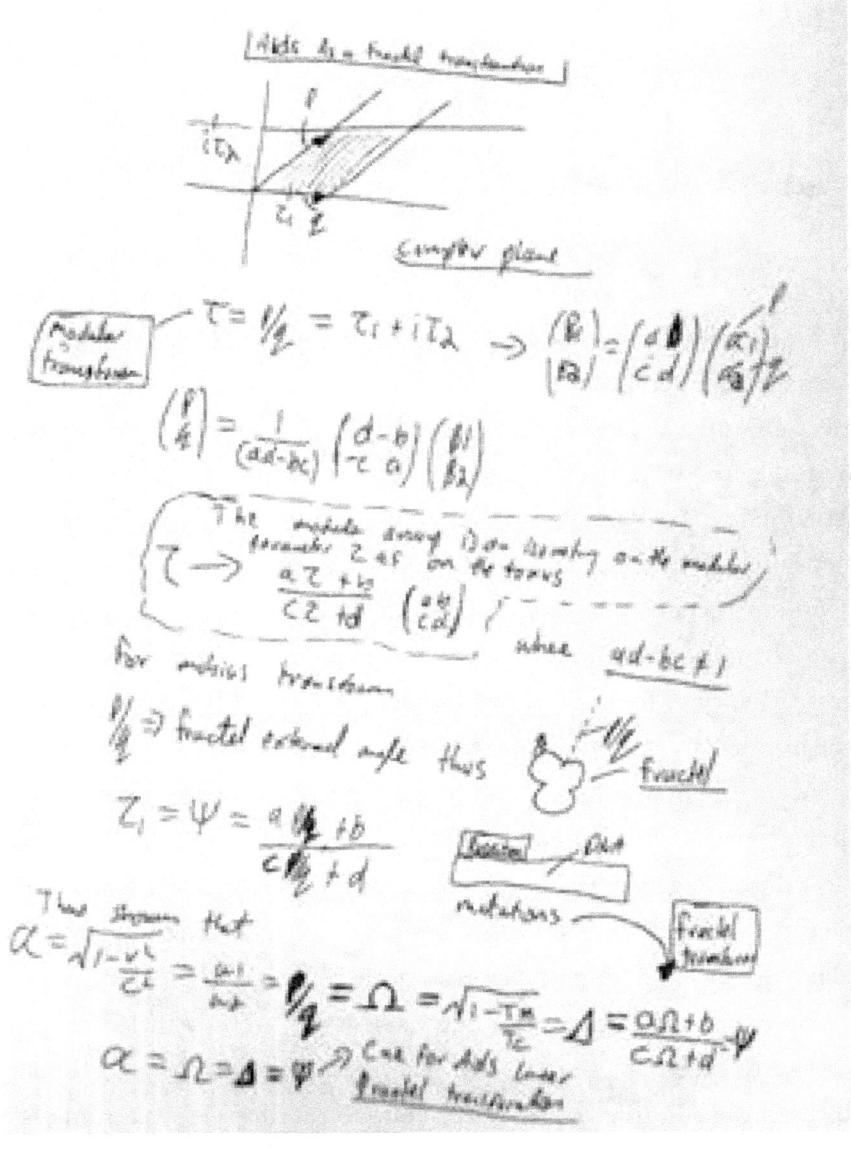

Fig 1 proof of the fourth equation

The previous diagrams put in the back of my previous book are the proof that fractals are responsible for most viruses and aids. The question is how do other dimensions play into this well if the universe is geometric and fractals form geometry then it's the fractals external angles that connect the different sections of a fractal together then Mobius transforms the Lorentz transform itself into an alternate reality because Mobius bands flip the constants but preserve triangles and circles and connect each entangled set. So the adjacent strands of DNA are connected by fractal external angles. In the book entitled magnetic monopoles by Ya Shnir there is further proof on page seventy-six and seventy seven that the complex number in the Mobius transform is indeed the winding number on a torus or circle map. Leading us to believe there is a connection between frames of reference. This is a Mobius transform and that this is what makes up the interdimensionality of reality. The adjacent strands on DNA are part of this fractal map between entangled pairs along the DNA strand. Put it this way aids is transforming dna to mix with its own RNA and this means its linear fractal transform on omega a transform on a transform. Which is exactly like the equation psi.

Since we know that DNA is a strange attractor around the pi bonds that are the center of DNA suggesting that the genomes are oscillating at initial conditions as depicted by my picture right here.

DNA

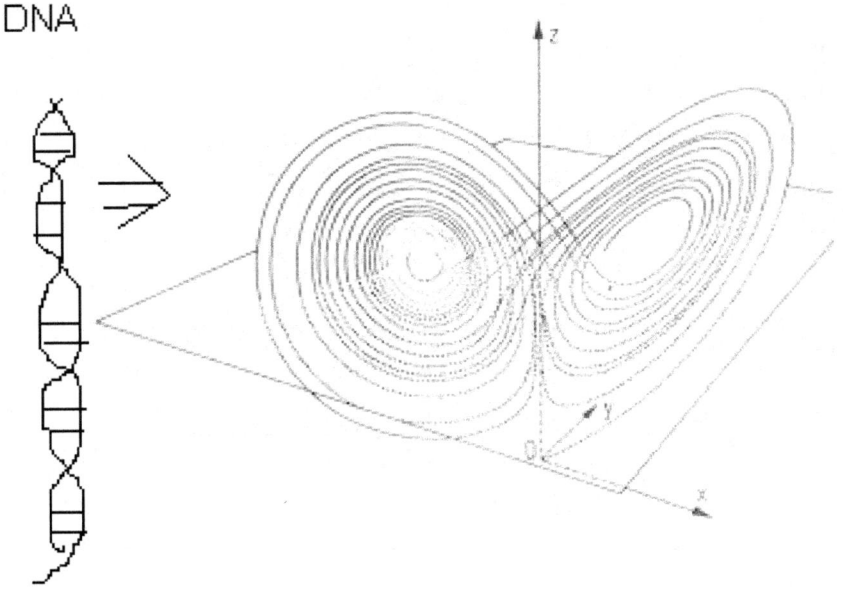

Fig. 2 DNA as a Strange Attractor.

This excerpt from Magnetic Monopoles by Ya Shnir proves that the Mobius transform is transforming omega on the torus which is proof for the fourth equation. One can say if the Mobius transform goes from one system of reference to another then if it's equivalent to tunneling and your frame of reference defines your universe or reality then the Mobius transform is a bridge to another reality. Thus there is a transformation between system of reference or winds and therefore one can change its position in a gravitational field instantaneously or the adjacent strand of DNA are physically linked it is changing the links for DNA. On the fourth equation it's possible for DNA to be orientated to Mobius strips which prove the fourth equation Mobius transform. In 1827 **Mobius realized** that a fourth **dimension would** allow a **three-dimensional** form to be **rotated** onto **its mirror-image**. $(p,-q)$ torus knot is the obverse (mirror image) of the (p,q) torus knot.[3] The $(-p,-q)$ torus knot is equivalent to the (p,q) torus knot except for the reversed orientation. Remember if alpha equal omega is the proof that tunneling and the pi bonds are quantumly entangled and there is infinite variation in finite space then between the pi bonds are connected linear fractal transformations since fractals are infinite variations in finite space. The frequency ratio w2/w1 from the diagram is equivalent to tau in his book w1/w2=tau=Ω=a((w1'/w2')+b)/(c(w1'/w2')+d). Further this is a transform form one system of another to change the velocity with no acceleration, which is very important.

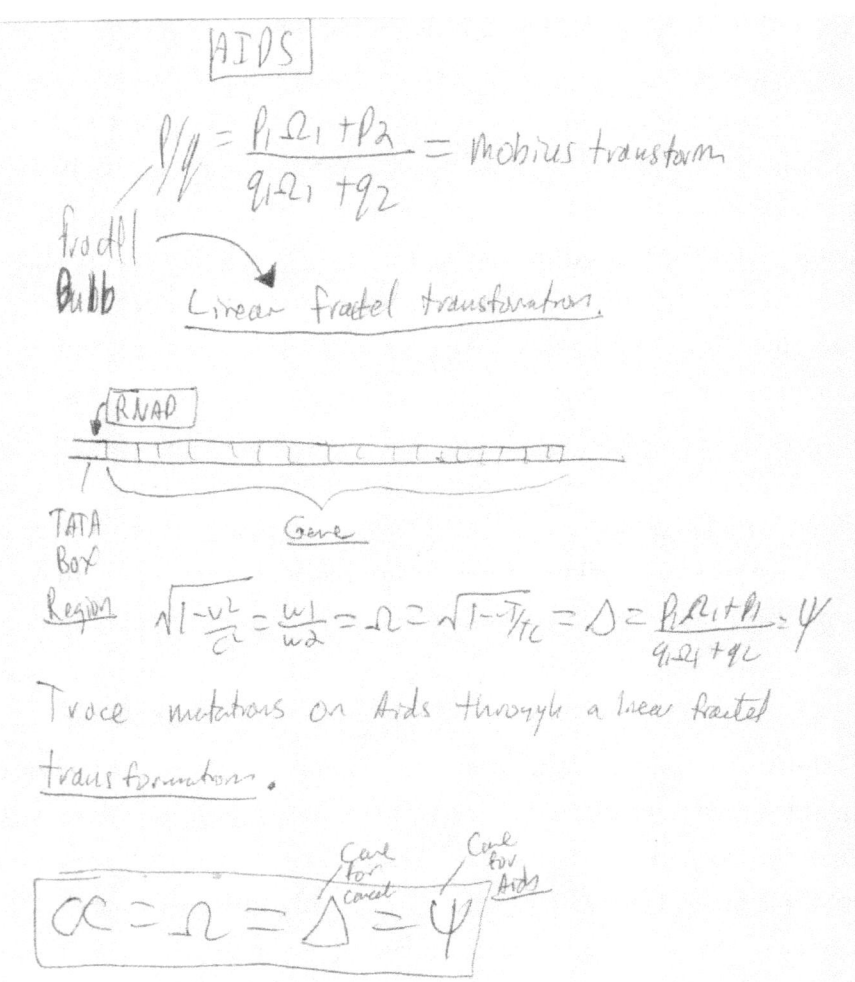

$$\frac{p}{q} = \frac{p_1 \Omega_1 + p_2}{q_1 \Omega_1 + q_2} = \text{mobius transform}$$

Frodel Bulb

Linear fractel transformation.

TATA Box Region

Gene

$$\sqrt{1 - \frac{v^2}{c^2}} = \frac{w_1}{w_2} = \Omega = \sqrt{1 - \frac{T}{T_C}} = \Delta = \frac{p_1 \Omega_1 + p_1}{q_1 \Omega_1 + q_1} = \psi$$

Trace mutations on Aids through a linear fractel transformation.

$$\alpha = \Omega = \Delta = \psi$$

Care for Cancer   Care for Aids

Fig. 4 The above Mobius transform is the form of a Farey tree between omega values p/q=(p+p')/(q+q') is in between windings numbers and thus is further proof for the fourth equation being equivalent to a circle map.

We see that the individual torus's form by the ring above and below the hexonogal carbon chains are at the center of those loops within the strange attractor.

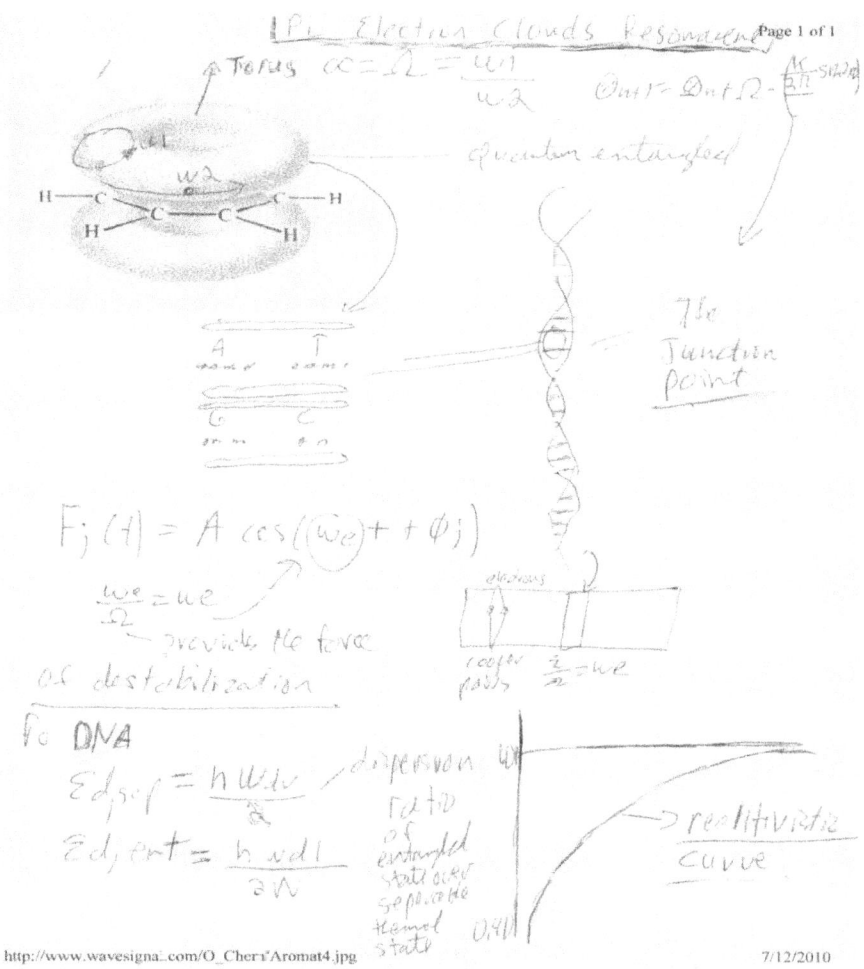

Figure 5 Electromagnetic torus's above the plane of the pi bonds proving entanglement.

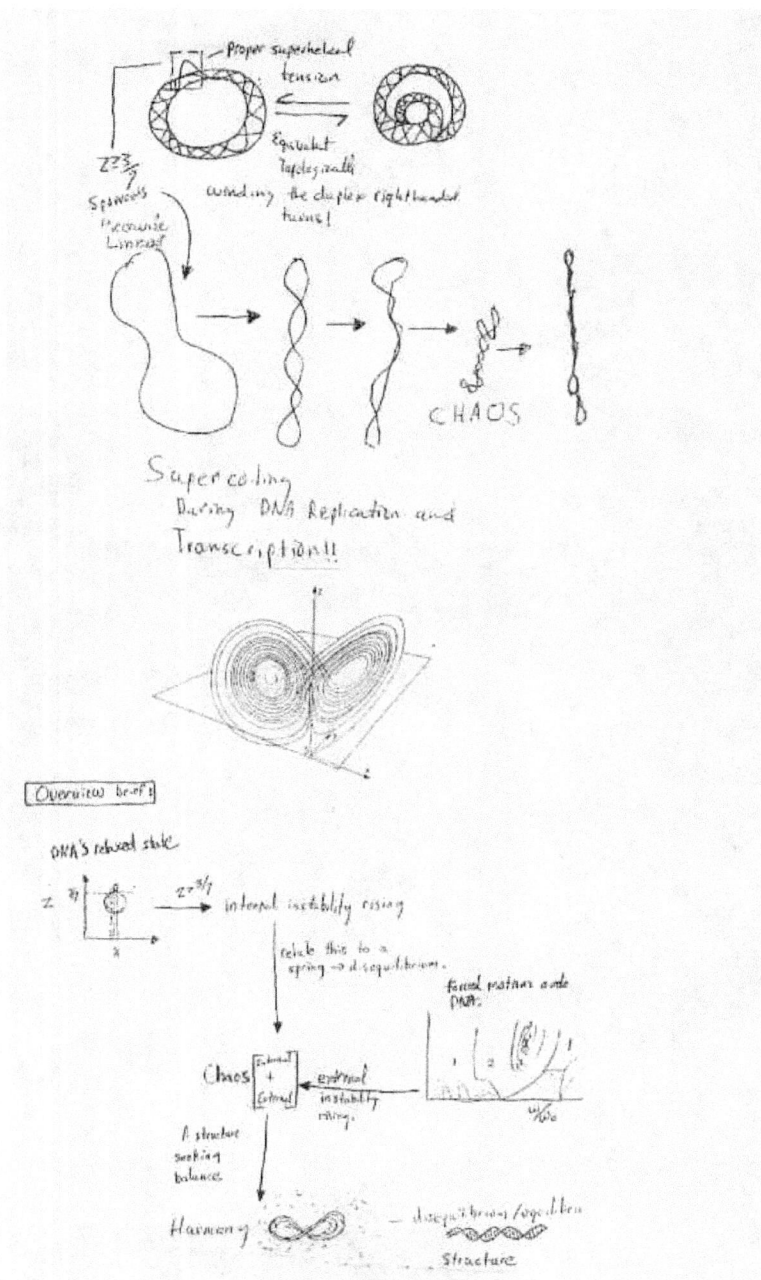

Figure. 6 Theory of DNA being Harmony structure in chaos.

Since each new torus is connected by a Mobius transformation we may say we are changing the frame of reference with Mobius transformations I suggest this how the adjacent strands of pi bonds are connected interdimensionally on DNA the sum of genetic information is thus Mobius transforms on omega winding numbers for the pi bonds and this is fractal thus we have shown that DNA is fractal.

Furthermore since alpha is being acted upon to go to a new alpha we can say that we are teleporting between frames of reference. Thus the forth equation is the teleportation equation. Thus since the two equations $E/\Omega=E^1$ $E\Omega=E^1$ and thus alpha is equal to omega and $E1/2KbTD$ across the junction $\alpha=\Omega=\Delta$ proves it's an inelastic collision across the superconducting junction then is Mobius transform from one inertial system before the junction to the one after words and we have grand unification and quantum teleportation. The lorentz trasnform determines your system of reference and this relates to quantum harmonic oscillators which determines uncertainty which is chaos and this relates to a mobius transform which changes your lorentz frame then it changes your universe since systems are relative. it transports you between systems of reference proving the universe is fractal.

Chapter 6

Nuclear Fusion

It makes sense that if two particles are colliding to give of heat as they stick together as in Einstien's relativity it also makes sense that if this is happening to cooper pairs it can be happening to nuclei of atoms when they stick together when fusion is happening the physics is in fact very simple and lets review that. If things are colliding on the torus they release heat. In order to accomplish nuclear fusion, the particles involved must first overcome the electric repulsion to get close enough for the attractive nuclear strong force to take over to fuse the particles. This requires extremely high temperatures, if temperature alone is considered in the process. In the case of the proton cycle in stars, this barrier is penetrated by tunneling, allowing the process to proceed at lower temperatures than that which would be required at pressures attainable in the laboratory.

Considering the barrier to be the electric potential energy of two point charges (e.g., protons), the energy required to reach a separation r is given by where $k=e^2/r$ = Coulomb's constant and e is the proton charge.

Given the radius r at which the nuclear attractive force becomes dominant, the temperature necessary to raise the average thermal energy to that point can be calculated.

Bottom line is that omega an inelastic collision is equal to delta quantum tunneling this is a new theory of fusion.

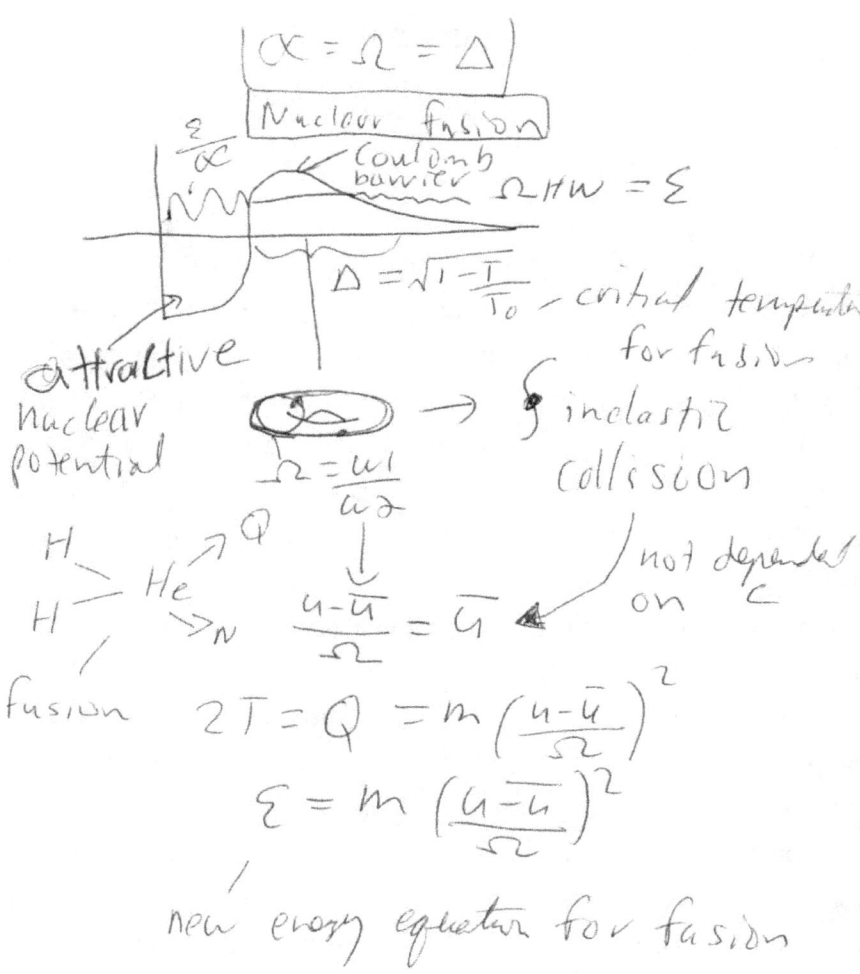

Fig. 1 Fusion across the coulomb barrier is an inelastic collision.

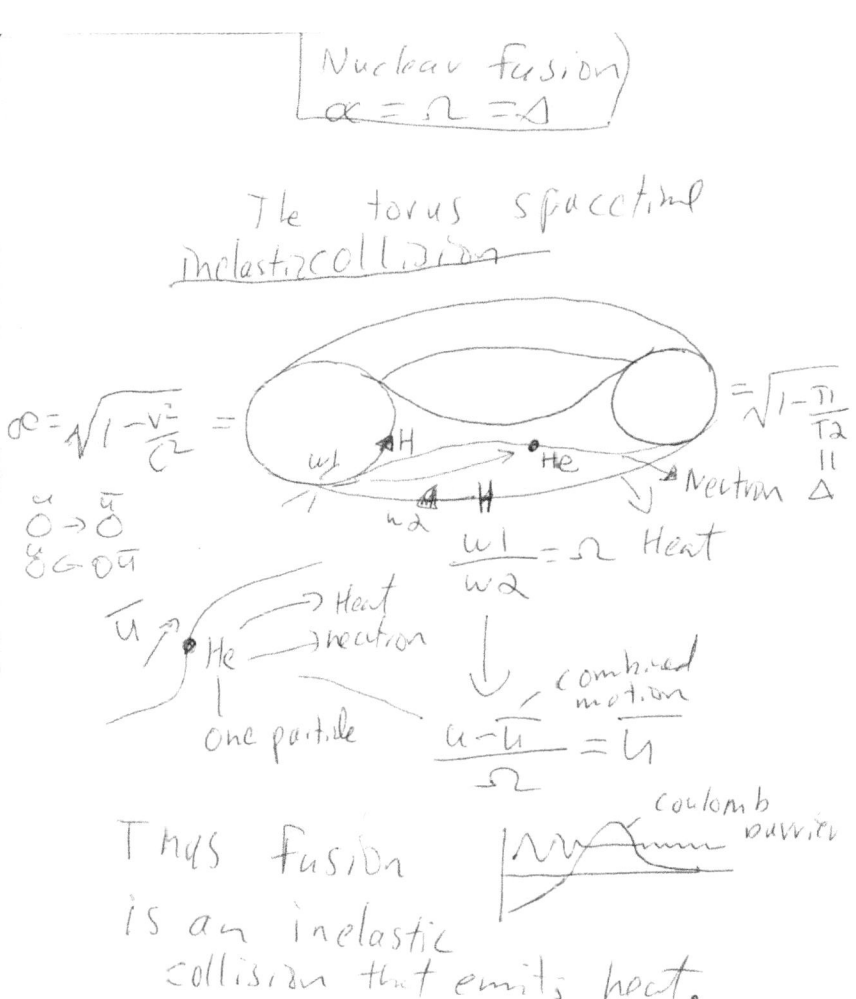

Nuclear fusion)
$$\alpha = \Omega = \Delta$$

The torus spacetime
inelastic collision

$$\infty = \sqrt{1 - \frac{v^2}{c^2}} = \qquad = \sqrt{1 - \frac{T_1}{T_2}}$$

$$\frac{w_1}{w_2} = \Omega \quad \text{Heat}$$

$$\frac{u - \bar{u}}{\Omega} = u$$

coulomb barrier

Thus fusion
is an inelastic
collision that emits heat.

Fig. 2 Torus and fusion.

69

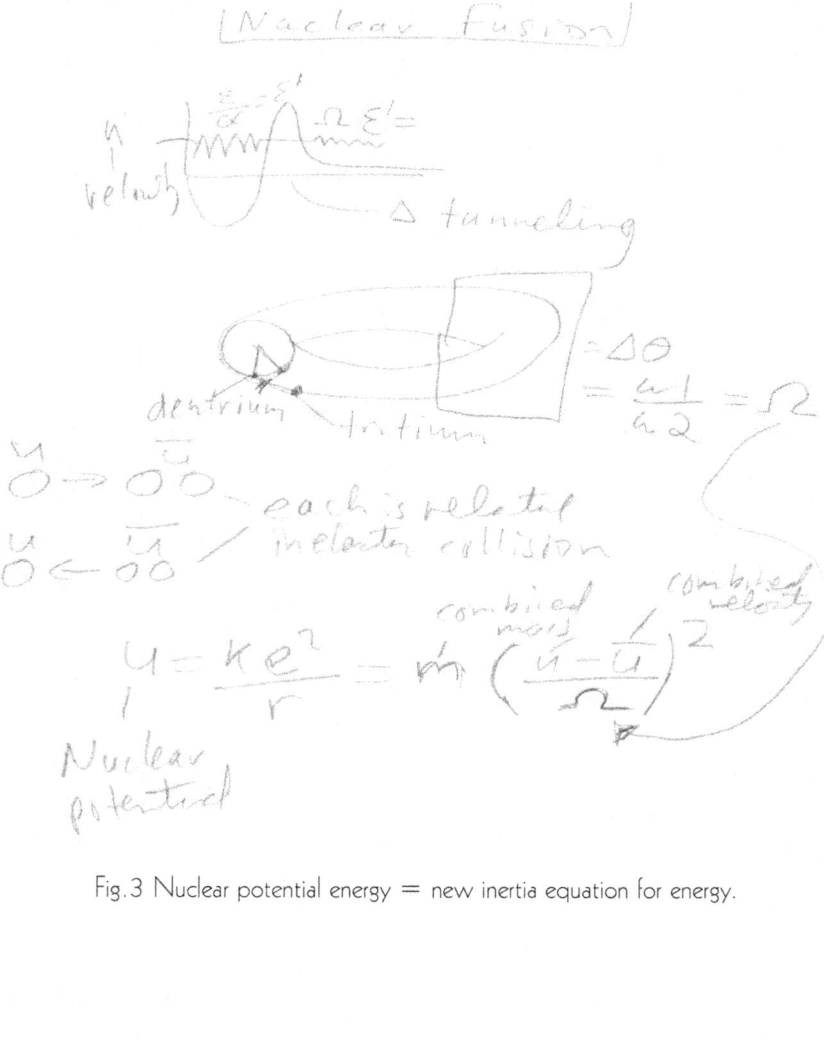

Evan Olsen

Fig. 3 Nuclear potential energy $=$ new inertia equation for energy.

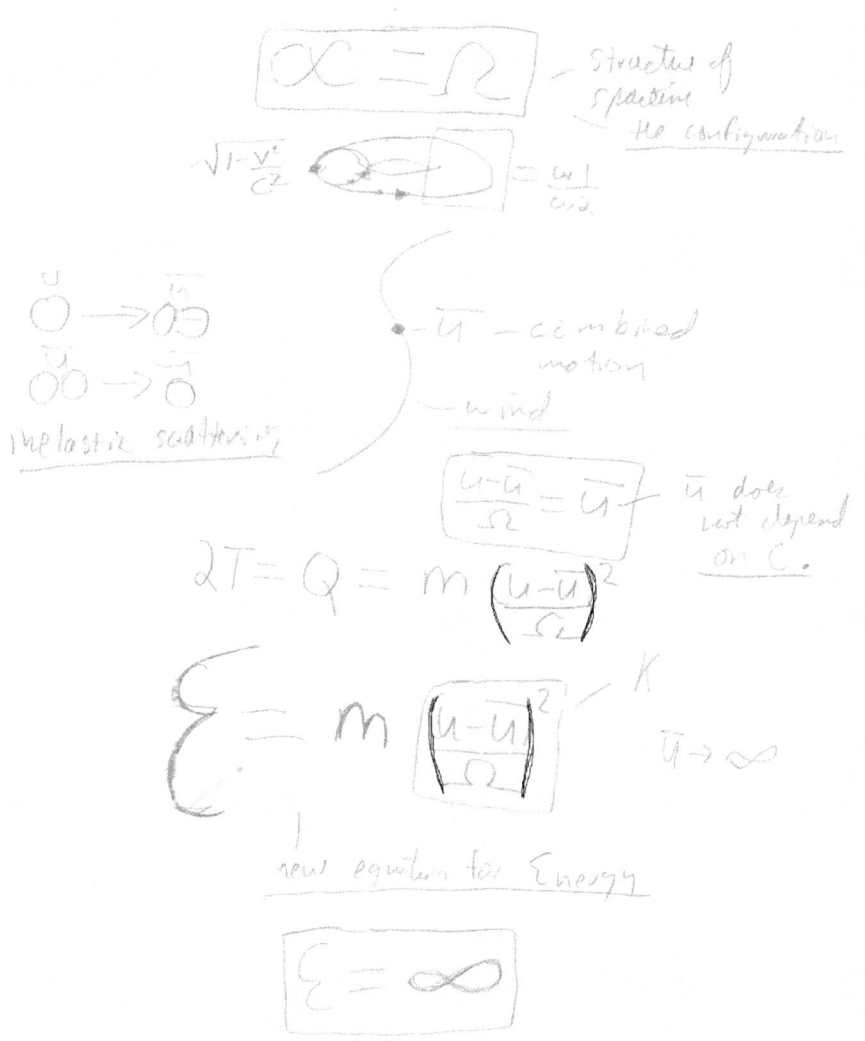

Fig 4 New equation for inertial energy.

The above equation for energy is in fact the equation a nuclear fusion reaction with inelastic scattering of heat due to a collision not based on constant of light. The key here is the ratio for omega which amplifies the kinetic energy of the particle after the collision.

Telepathy and the space time continuum.

Fractal and Chaotic Dynamics in Nervous Systems

Chris C. King

Department of Mathematics & Statistics,

University of Auckland.

Abstract shows that $\Omega = \tau s /\tau$ [4.13]

where ts is the interstimulus distance and tau is natural period. Periodic stimulation of giant squid axons have similarly demonstrated mode-locking, bifurcation to chaos and the Transmission of chaotic signals along the nerve axon. If the alpha equal omega delta equation part is true then space-time can permeate consciousness and neurons are linked by quantum tunneling and thus the universe is a consciousness if space-time is linked by interconnected circle maps. The brain uses circle maps in describing neural firing then it simulates the structure of the space time continuum.in fact if the space-time continuum functions like neural nets and are fractal then you would expect the superstructure of the universe to resemble a neural net and if one look at the pictures it does indeed this is fractal emergence through chaos. The universe is a giant neural net and consciousness can permeate through time and space. As we can see here 125 million parsecs the universe resembles a neural net with clumps resembling neurons and the dendrites stretching out.These pictures prove the brain is fractal and so is the universe.

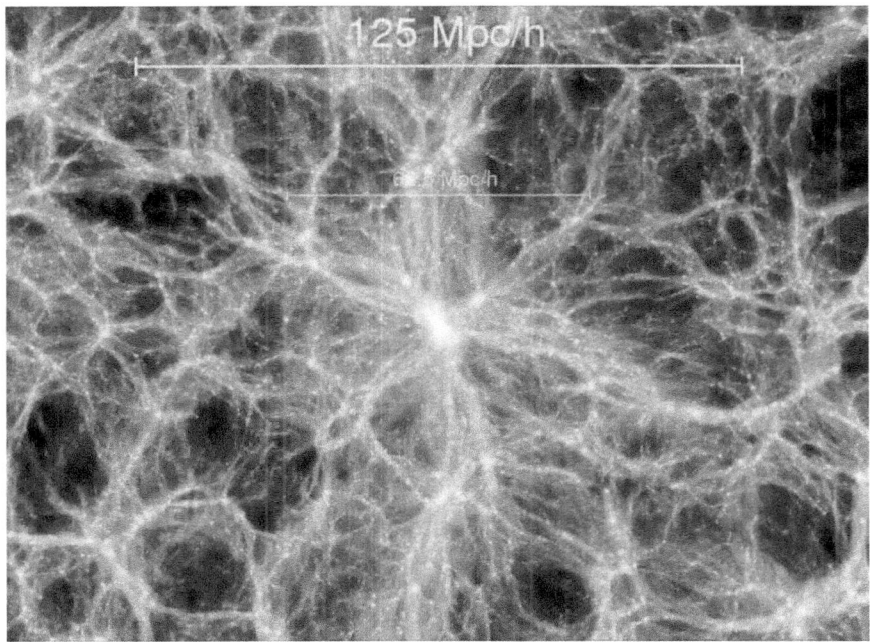

And tau is the neural period

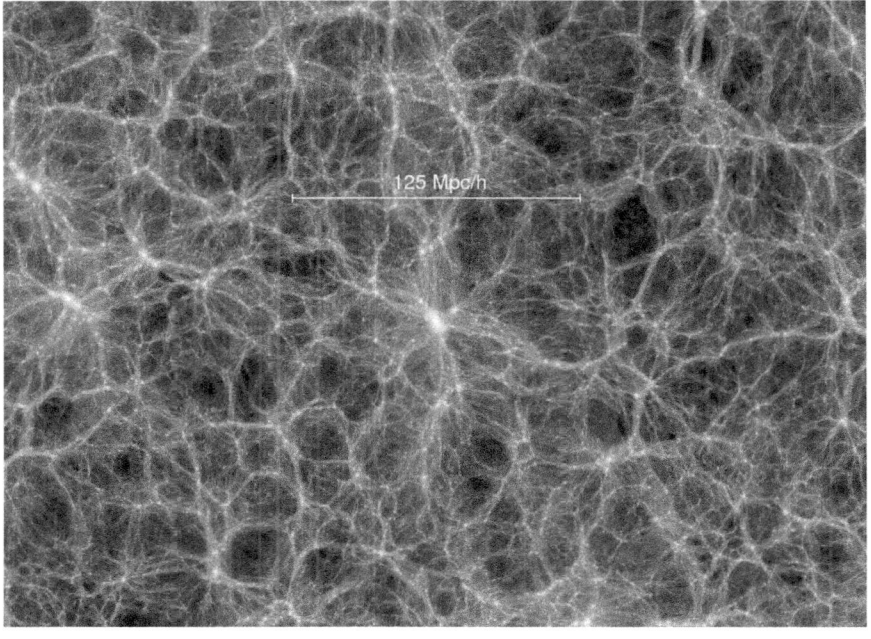

Here is a neural net

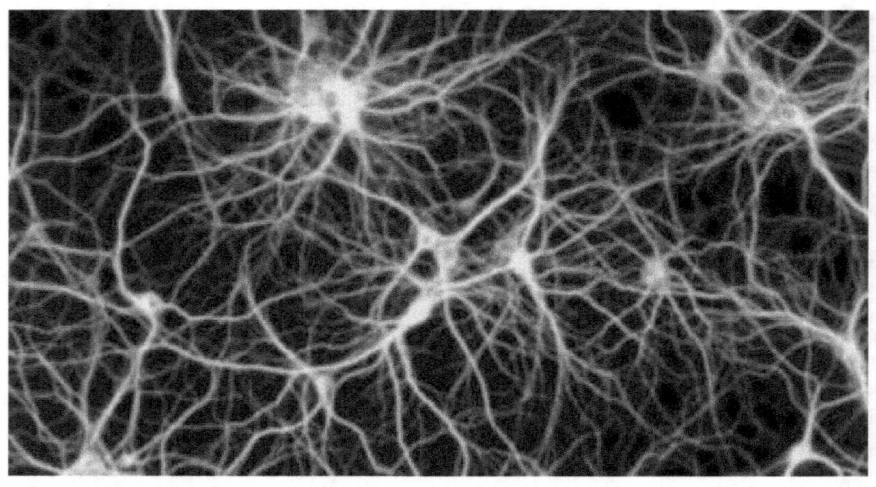

www.ingramcontent.com/pod-product-compliance
Lightning Source LLC
Chambersburg PA
CBHW061515180526
45171CB00001B/195